The Chora of Metaponto 2

The Chora of Metaponto 2

Archaeozoology at Pantanello and Five Other Sites

Studies by
Sándor Bökönyi and Erika Gál

Edited by László Bartosiewicz

Institute of Classical Archaeology
The University of Texas Press

Printed in the United States of America

First edition, 2010

Requests for permission to reproduce material from this work should be sent to:
Permissions
University of Texas Press
P.O. Box 7819
Austin, TX 78713-7819
www.utexas.edu/utpress/about/bpermission.html

∞ The paper used in this book meets the minimum requirements of
ANSI/NISO Z39.48-1992 (R1997) (Permanence of Paper).

Library of Congress Cataloging-in-Publication Data

Bökönyi, Sándor.
The chora of Metaponto 2: Archaeozoology at Pantanello and five other sites / studies by Sándor Bökönyi and Erika Gál ; edited by László Bartosiewicz.
p. cm.
Includes bibliographical references and index.
ISBN 978-0-292-72134-0 (cloth : alk. paper)
1. Metapontum (Extinct city) 2. Pantanello Necropolis Site (Italy) 3. Historic sites—Italy—Metaponto Region. 4. Excavations (Archaeology)—Italy—Metaponto Region. 5. Animal remains (Archaeology)—Italy—Metaponto Region. 6. Animal culture—Italy—Metaponto Region—History. 7. Metaponto Region (Italy)—Antiquities. I. Gál, Erika. II. Bartosiewicz, L. III. University of Texas at Austin. Institute of Classical Archaeology. IV. Title. V. Title: Archaeozoology at Pantanello and five other sites.
DG70.M52B65 2009
937'.773—dc22

2009021211

For reasons of economy and speed, this volume has been printed from camera-ready copy furnished by ICA, which assumes full responsibility for its contents.

Cover and title page illustration: Sacrifice scene from an Attic black figure lekythos, attributed to the Leagros Group. Tomb 292-1, Pantanello Necropolis, 1986 ICA excavation. Rendering by Sandra Langston.

Dedicated to the memory of
Sándor Bökönyi
and
to the new generation
of Hungarian archaeozoologists.

Photo: Metaponto, 1991 (CW)

Special Acknowledgment

This is the second of a projected multi-volume series on the chora of Metaponto. The first volume, *The Chora of Metaponto: The Necropoleis,* was published in 1998, and for a while it seemed it might be the last. This renewed publication effort could not have been conceived, much less implemented, without the active and generous support of the Packard Humanities Institute (PHI).

One of the greatest needs in Classical Archaeology today is the dissemination of primary research to both the scholarly community and the wider public. Collected and studied by modern interdisciplinary methods, the data generated from this research will yield a harvest of new and original contributions to the field. The need is all the more urgent as urbanization and agriculture rapidly obliterate the ancient landscape, the source of new knowledge. Because of the increasingly complex technological nature of modern archaeological investigation, as well as the sheer mass of data involved, presentation must be carefully organized and amply illustrated. That is the goal of this series.

The Packard Humanities Institute has recognized these needs. In 1996, when the archaeology publication program of the National Endowment for the Humanities was drastically curtailed, it seemed that there was no organization with the resources—and historical vision—to make possible large scale research projects and their publications. PHI answered the prayers of ICA and other important projects throughout the Classical world. We express here our overriding debt of gratitude to PHI for this volume and those to come.

Contents

Acknowledgments

Archaeological excavations carried out annually from 1974 to 1991 by The Institute of Classical Archaeology (ICA) in the territory (*chora*) of Metaponto provided much of the material basis for this study of the ancient fauna. We are grateful to the generations of students who contributed their labor, thought, and enthusiasm to the enterprise, and to those private donors and foundations, especially the James R. Dougherty Jr. Foundation and the Brown Foundation, who made it financially possible. For much of this period—nearly two decades of sustained and uninterrupted effort—the National Endowment for the Humanities provided vital grants of matching funds. The students and many of these donors took particular interest in the archaeobotanical and archaeozoological aspects of the work, which at this time and in this region of the world was a pioneering contribution without parallel. Among the students and colleagues whose names are recorded in ICA's annual reports for these years. I would like to single out that of Lorenzo Costantini, whose laboratory of Bioarcheologia in Rome carried out not only the paleobotanical analyses from ICA's excavations, but also the first studies of the faunal remains. Among the participants in the excavations in 1980 and 1981 was a young student of paleontology at the University of Texas, Boyce Cabaniss, who subsequently followed a career in law, but whose early interest is still keen and who kindly proofread the manuscript of this volume.

A necessary condition for the success of a project of the scope and duration of the Metaponto excavations was the close and cordial working relationship with the Italian authorities. Not only was the path smoothed by the Soprintendenza Archeologica della Basilicata, but an atmosphere of collaboration and intellectual engagement was also present from the beginning, first under the inspiring leadership of the first Soprintendente for the Region of Basilicata, Professor Dinu Adamesteanu, and then under his successor. Adamesteanu's generosity and charisma drew archaeologists and scholars from many nations—France, Germany, Romania, and Hungary—to cite just a few of the groups who have worked harmoniously together between the Bradano and the Sinni, and continue to do so to the present. A special thanks goes to Dr. Antonio De Siena, the Director of the Museo Nazionale di Metaponto, for his generosity not only in providing ICA's archaeological mission with unstinting assistance and ideal working spaces, but for his active support and interest in the research. Thanks to him, Professor Bökönyi was able to study the fauna of Bronze Age Termitito, and that from the excavation of the drain (*cloaca*) in the ancient city sanctuary. The spirit of international collaboration, so enthusiastically promoted by Adamesteanu, lives on and continues to make Basilicata a constantly exciting and rewarding place to carry out research.

At his death on Christmas Day, 1994, Sándor Bökönyi's work was suddenly left to others to complete. We are grateful that it came into the competent hands of Dr. László Bartosiewicz, Dr. Erika Gál, Éva Nyerges, and Anna Biller, who responded to the challenge with exemplary professionalism and promptness. It has been a pleasure to work with them.

The contributions of these formidable groups of scholars, which have added substantially to the range and scope of the original work, have been further enhanced by the presentation skills of yet another team: production editor and graphic artist Deena Berg, a good archaeologist in her own right; Chris Williams, series designer; and Jessica Nowlin, whose GIS skills helped generate our maps. Erika Gál took all black and white photographs of bone materials. Color photographs were taken by Chris Williams, Alberto Prieto, and Cesare Raho. Anna Biller created the black and white drawings.

All of these contributions and the production of the volume you hold in your hands would not have been possible without the generous and constant support of the Packard Humanities Institute.

Joseph Coleman Carter
Pantanello
May 2008

Introduction

I. The Animals and Their Chora

Joseph Coleman Carter

This volume of archaeozoological studies marks the renewal of a major program to document more than thirty years of research in the chora of Metaponto *(Fig. I)*. Bringing this book to print has been made possible by the generous ongoing support of the Packard Humanities Institute. This study and those to follow are the first serious attempts to deal with the ancient rural economy in the context of Classical Archaeology in South Italy. Despite the tradition of such research by prehistorians, faunal reports even from prehistoric sites have not been standard in this geographical area. With this publication, the Institute of Classical Archaeology (ICA) begins to fill a serious gap in our knowledge of ancient Italy, and hopes to set a precedent for similar studies.

The archaeozoological analysis, an integral part of the studies of the chora of Metaponto, receives the pride of place at this time because of a special set of circumstances. The full publications of the archaeological sites from which these materials originate are underway and will appear soon and at regular intervals in the coming years, but Sándor Bökönyi's study of the fauna from the chora was essentially complete in 1989. There was much more that he would have wished to do, but the main results had already been reached before his work was cut short by his premature death in 1994. Fortunately, his younger colleague, László Bartosiewicz, rose to the challenge of completing the project, and along with his students—Erika Gál, Eva Nyerges, and Anna Biller—brought this book to its present form. Each made their individual contributions, carefully editing, annotating, and illustrating Bökönyi's original manuscript. As a result, this volume is the first of the projected stud-

Figure I Chora of Metaponto. View of a valley in the Tempe Rosse locality between the Bradano and Basento, looking northwest (AP).

ies and will take its place in the series, *The Chora of Metaponto,* after the initial two-volume study, *The Necropoleis* (Carter 1998a), which was published a decade ago. Subsequent volumes will include the farmsites from the chora—several of which furnished important material for this work—and the major sites of Incoronata and Pantanello. The latter, with its Greek sanctuary, Roman tile factory and Late Neolithic site, produced the bulk of the evidence for the fauna of the chora of Metaponto. The dates given here are based on our current understanding of the sites, although further refinement is expected when the final analyses of the ceramic materials—particular the Pantanello complex—are completed.

From early on, the research that ICA began in 1974 was interdisciplinary in nature. How can a countryside or any ancient settlement be fully understood without comprehending the plants and the animals—both wild and domesticated—that populated it? First came the study of plant remains in 1978, when Lorenzo Costantini began to analyze the extraordinary material emerging from the Greek sanctuary at Pantanello. That project led directly to the study of faunal remains and the participation of Sándor Bökönyi. What followed was a study of the human remains from the Pantanello Necropoleis (and two other burial sites) by Renata and Maciej Henneberg, who eventually analyzed the burials in the urban necropolis and elsewhere in the chora. Subsequent volumes in *The Chora of Metaponto* series will present these palaeobotanical and physical anthropological studies.

While there were delays in finding trained personnel to analyze the skeletal remains, bone material was systematically recovered at every phase of the excavations, beginning with the first excavation at Pantanello in 1974. The animal and human bone material was recorded with the same attention to context as the ceramics and other archaeological materials. This

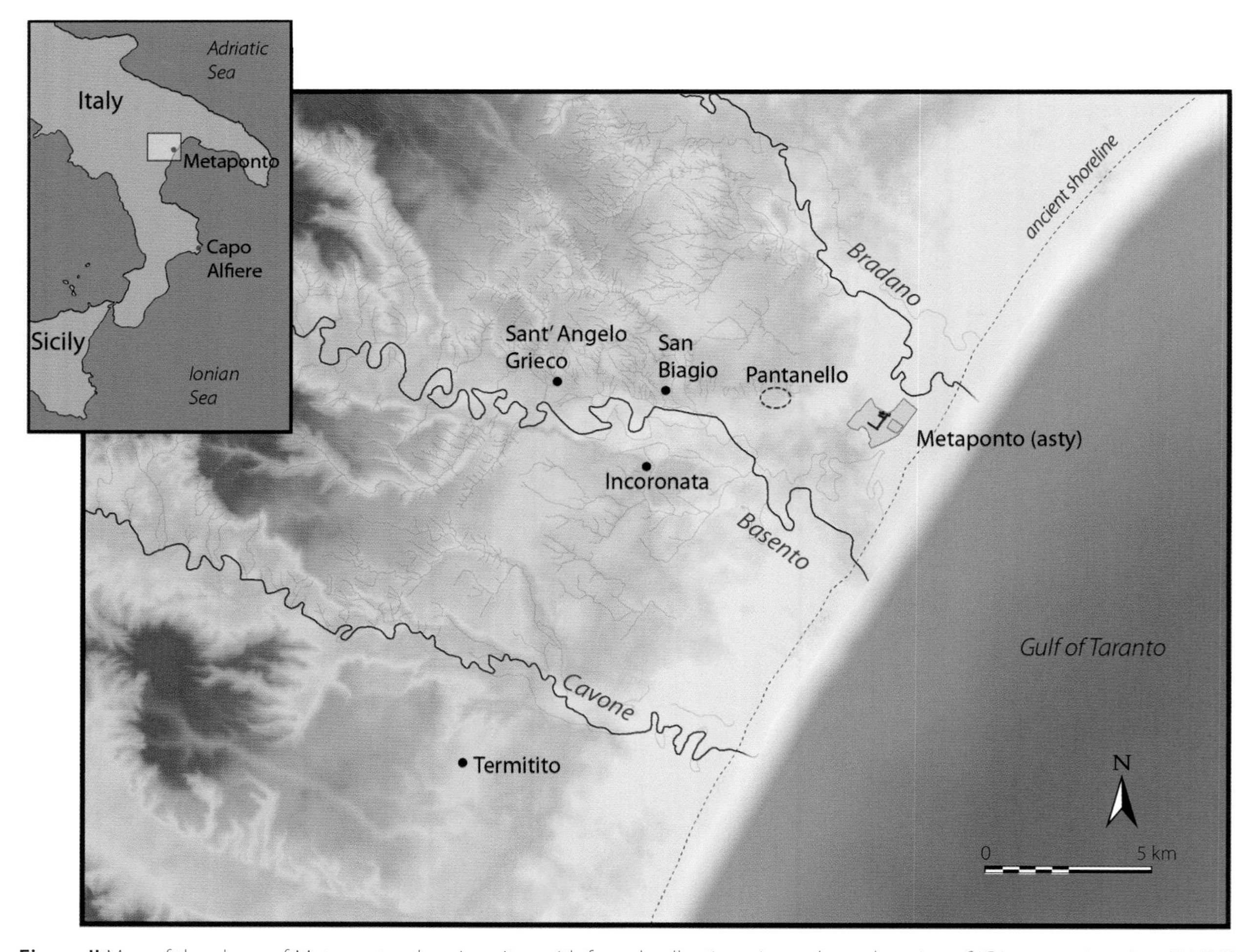

Figure II Map of the chora of Metaponto, showing sites with faunal collections. Inset shows location of ICA excavation sites (JN/DB).

Site	Date
Pantanello Pits	Late Neolithic
Termitito	Late Bronze Age
Incoronata	8th–6th c. BC
Pantanello Complex: Greek and Roman Periods	6th c. BC–3rd c. AD
Greek Pit (Mixed Deposit)	6th c. BC–3rd c. AD
Pantanello Sanctuary	6th c. BC–3rd c. BC
Pantanello Necropolis	6th c. BC–3rd c. BC
Pantanello Kiln Deposit	2nd c. BC–1st c. AD
Metaponto Sanctuary	4th c. BC
Sant'Angelo Grieco	6th c. BC–1st c. AD (primarily 2nd c. BC–1st c. AD)
San Biagio	Late 3rd –4th c. AD

Table I Sites in the Chora of Metaponto.

procedure resulted eventually in large representative collections of precisely dated faunal material from all of the sites excavated by ICA in the chora. Careful recording was also practiced by the Soprintendenza for two additional collections—the Late Bronze Age site at Termitito and the drain (*cloaca*) at the Metaponto Sanctuary—that add new temporal and spatial dimensions to this study. Unfortunately, the constraints of time and resources permitted only limited screening for fine bones at a few sites.

Figure III Aerial photograph of Pantanello. *Foreground,* Neolithic Pits; *left,* Greek Sanctuary; *right,* Roman tile factory and Kiln Deposit (A. La Capra/ICA).

Six sites furnished the material for this volume *(Fig. II; Table I).* These go as far back as the 3rd millennium BC, beginning with the Late Neolithic site at Pantanello, a village overlooking the Ionian coastal plain and the Basento River *(Fig. III).* Twenty-six pits containing ceramics, fine lithic tools and obsidian, some evidence of copper metallurgy, and a significant faunal collection, were excavated in 1983 and will be the subject of a separate volume in this series.

Bökönyi was also able to examine the faunal collection from the Late Bronze Age site of Termitito *(Fig. IV),* thanks to the generosity of its excavator, the current Director of the Museo Nazionale di Metaponto, Antonio De Siena, who currently dates the deposit to the 13th–12th century BC. The animal bones were part of the fill of a large and deep pit, the substructure of a semi-interred residence and probable administrative center, which crowned a high plateau inland from the coast and immediately to the south of Cavone

Figure IV Aerial photograph of Termitito, with Late Bronze Age site in foreground (courtesy of A. De Siena).

River and the historic chora of Metaponto. Although some information is available in scholarly articles, a full publication of the site is eagerly awaited.

As will be shown in this study, major change in the chora was brought about by the arrival of Greek colonists, who were accompanied by some new as well as superior forms of livestock. This shift is uniquely documented at the site of Incoronata, an Iron Age indigenous village site located inland on a plateau, which had already experienced strong Greek influence in the 7th century and was subsequently occupied

Figure V Aerial photograph of the Iron Age site at Incoronata from the southeast, looking inland (Courtesy of P. Orlandini).

by a colonial sanctuary in the middle of the 6th century BC *(Fig. V)*.

The bulk of the material analyzed in the present study was excavated over the course of a dozen campaigns between 1974 and 1991 at the site known as Pantanello *(Fig. III)*. This area actually comprises a complex of partially overlapping sites from different periods. The first is the above mentioned Late Neolithic site, overlooking the Basento River to the south on the last and lowest marine terrace before the coastal plain. The second is the rural Greek Sanctuary, active from the 6th century through the 3rd century BC and located on the south slope of the terrace *(Fig. VI-VII)*, which produced a wealth of botanical material as well as faunal remains around a spring. Contemporary with this was the third site, the Pantanello Necropoleis, located to the west *(Fig. VIII)*. The Tile Factory and related Kiln Deposit, just to the west of the prehistoric site, constitute a fourth site. Dating to the 2nd century BC–1st century AD, this area produced the bulk of the animal remains. A fifth location, the Greek Pit, overlaid the Neolithic pits, but the mixed materials could not be dated to a specific period.

Figure VI Pantanello Sanctuary and spring, lower terrace (CW).

Also thanks to Antonio De Siena, Bökönyi was able to include in his study animal remains from the 4th century BC drain (*cloaca*) at the Metaponto Sanctuary *(Fig. IX)*. More faunal remains from the period when Rome dominated the territory of the former colony were also collected at two other sites: the large late Republican, early Imperial farmhouse at Sant'Angelo Grieco *(Fig. X)* and the small Late Roman villa at San Biagio *(Fig. XI)*, both excavated in the early 1980s (see Carter 2006 for summary discussion of these sites).

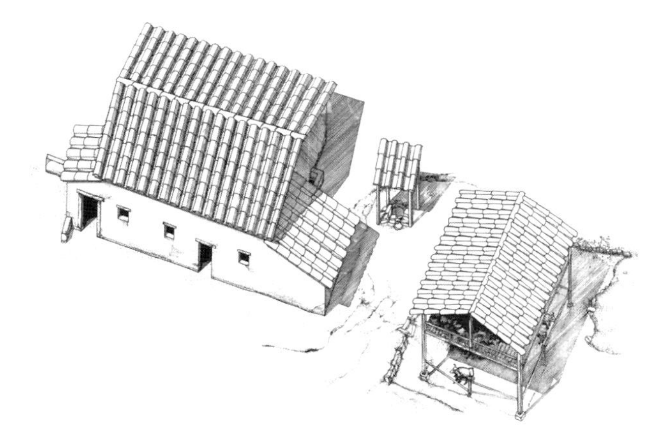

Figure VII Reconstruction of the farmhouse built on top of the earlier oikos at the Pantanello Sanctuary. A collecting basin that provided water for animals in the 6th century BC was restored to use in the second half of the 4th century BC (A. Patterson/ICA).

Archaeological investigation of the fauna from the chora of Metaponto has added a new dimension to our knowledge of the ancient countryside by opening up a window on the environment and the ancient economy. Nothing like the present study exists for any other area of southern Italy and for few other areas of the ancient Mediterranean. This is a pioneering effort, and it is indeed fortunate that it was first attempted by a scholar of Bökönyi's caliber. Up to this point, what was known of the animal species that inhabited this territory and most ancient territories was limited to the species mentioned by the ancient authors, and this was specialized knowledge contained in agricultural

Figure VIII The Pantanello Necropoleis (A. La Capra).

Figure X The site of Sant'Angelo Grieco (CW).

treatises, or the casual mention of animals in literary works (Douglas 1928; Thompson 1895). These writings are no substitute for a systematic, statistical, and diachronic study of the actual remains of a single area of the ancient world that covers all of the major periods from prehistory to the Middle Ages. We knew from Bacchylides (*Ode* XI.119–120) that the land beside the Basento was *hippotrophos* or "horse-nurturing." Now we know, from Bökönyi, that the Metapontine not only raised horses but bred them for size and riding, and we know how these horses compared with horses from many other areas of the ancient world even beyond Greece and Rome.

The ancient countryside comes alive when we learn that it was populated by now rare and extinct animals, such as ibex and the mighty aurochs. Black vultures, no longer found in Italy, soared in the sky over Metaponto, and their bones were perhaps used to create flutes (see Gál, Chapter 5). Such information satisfies a deep human curiosity to know about man's animal neighbors in the past. But this study also provides primary evidence for the environment, the physical context of human settlement and its changes over time, and for the major shifts in the agricultural pastoral economies of its human communities.

The area around Metaponto supported varied wild fauna, each with its own requirements for survival. It may be surprising to learn that deer were plentiful throughout the three millennia under study. They continued to be an occasional food source even dur-

Figure IX Aerial view of the Metaponto Sanctuary (courtesy of Dieter Mertens/Soprintendenza archeologica della Basilicata).

Figure XI The late Roman farmhouse at San Biagio. The natural terraces become larger closer to the shore (CW).

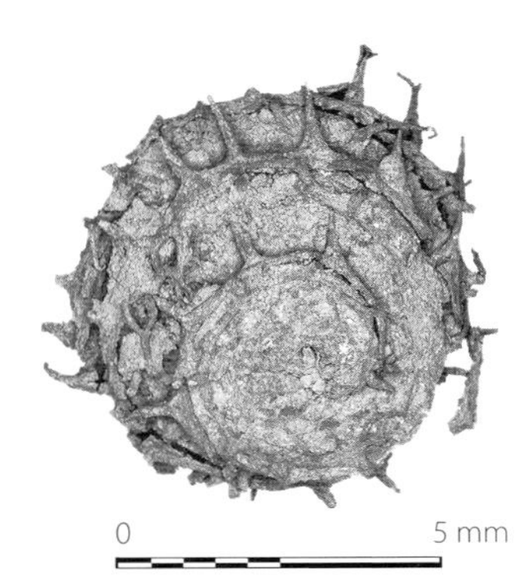

Figure XII The oldest known alfalfa seed in Italy, ca. 300 BC, Pantanello Sanctuary (L. Costantini).

ing the dense settlement of the chora and its intense cultivation by the Greeks. Hence there were forests in close proximity, and gallery forests probably along the streams. This finding supports that of the pollen record (Carter 2006, 28, Fig. 1.20). Forests of pine, maple, and oak existed throughout the colonial period, despite the demands for fuel to fire the kilns and metal working establishments that proliferated throughout the chora. The low-level of *macchia* (scrub) species (*Pistachia, Phillyrea*) in the Archaic period (until the 7th and 6th centuries BC), and their rise in the Classical period (5th and 4th centuries BC) makes sense when we learn that sheep and, especially, goats—browsers that thrived on these plants—were in the majority among domesticated animals in the early colonial period. These animals then lost ground to oxen, the "tractors" of the dominant plow-based agriculture that was the basis of Metapontine prosperity for three centuries or more from the second half of the 6th century onward. With the decline in the number of goats, the macchia was then able to make a temporary comeback.

It is interesting to learn that among "the firsts" in this detailed study of the fauna from a Greek site in southern Italy is the earliest known chicken on the peninsula. That these Greeks introduced this typical farmyard species in the west should not be a surprise. It might have happened anywhere along this coast or in Sicily, and perhaps, if similar studies should appear, we shall discover that it did.

Another historically significant find by Bökönyi is that the colonists were responsible for introducing a larger, superior breed of sheep (also identified at Incoronata in a 6th century BC context). These replaced the smaller sheep descended from Neolithic imports three or four millennia earlier. This opens possibilities for a wide range of reflections on the role animal species may have played in marginalizing the indigenous animal populations and human inhabitants of the Metapontine chora and surrounding areas. "Biological imperialism" was a major aspect in the colonization of the New World (Crosby 1986; Mann 2007). Investigations of the early English settlement at Jamestown, to cite one striking example, show how introduced species can profoundly change a world, albeit the cultural and biological differences between colonizers and colonized in that case were far greater than in this Mediterranean situation. This type of concrete evidence offers a valid alternative source, beyond the few and much debated passages from ancient authors, on the relations between Greek colonists and indigenous populations in their early years of contact.

One of the most original and important contributions of this study, particularly as it regards the Greek and Roman periods, is the discussion of animal husbandry and the conscious breeding of livestock for size. In his analysis of cattle and horse remains, Bökönyi brought to bear a global knowledge of these and other species. He concluded that the Metapontines were raising the largest known cattle in the ancient world in the 2nd century BC, a period that saw, in the famous judgment of no less an historian than Arnold Toynbee, the former Greek colonial areas of southern Italy reduced to a desert occupied only by sheep and slaves (Toynbee 1965). He admitted that the archaeological evidence at the time, the mid-1960s, was slight. Toynbee relied almost exclusively on ancient texts by Roman authors writing two centuries or more after the fact. The faunal remains, through the prism of Bökönyi's expert knowledge, tell another story.

In an equally unexpected way, the skeleton of a horse from a burial in the Pantanello Necropoleis illuminates another aspect of the history of Magna Graecia. As Bökönyi recognized, after measuring the bones, this was no ordinary equine specimen, a fact underscored by its careful burial in its entirety—much in contrast to the fate of other horses, whose remains, as attested by the Pantanello Kiln Deposit, show that they were exploited for raw materials. Much larger than the norm, this horse was not, in Bökönyi's first-hand knowledge of these breeds, a Greek, or a Roman, or even a Scythian horse, but most probably a horse of Persian descent. How did it reach Metaponto in southern Italy?

The most likely explanation is that the horse was brought from Greece by one of the Greek mercenaries (the so-called *condottieri*), many of whom, like Pyrrhus of Epirus, came from the west coast of Greece, called by the Italiote Greeks to help them in their struggles with Lucanians and then the Romans in the late 4th and early 3rd centuries BC. The probable date of the horse burial fits the time period, but what linked the horse and its "condottiero" was the horse's bloodlines—and its diet.

It is a little commented upon but striking fact that the booty from Alexander's conquest of the Persian Empire included the prized royal Nisean horses belonging to the Great King and his vassals further east, e.g., in the area of modern Turkmenistan (Arrian, *Anabasis* VIIa.13; Green 1991). The Nisean horses grew to their great size in part because they grazed on pastures of alfalfa—*medica* in Latin, from its Medean origin. This wonder crop was introduced to Greece from Persia, supposedly at the time of the Persian Wars in the early 5th century BC. Our earliest literary source for its use in Italy is the 1st century AD Roman agricultural writer, Columella, who singles this fodder out for praise: "*pabulorum optima sunt medica et faenum Graecum*" (*de Re Rustica* 2.7.2) The first identified physical evidence comes from a late 4th century BC deposit at the Pantanello Sanctuary, thanks to the paleobotanical investigation by Costantini *(Fig. XII)*. It may, on the one hand, be pure coincidence that the Persian horse and its favorite food appear at the same site at the same time. On the other hand, it might reflect the undeveloped state of biological research in Classical Archaeology. The evidence may be there, but trained eyes are needed to see it.

And so there is good reason to think that the occupant of Tomb 316 at Pantanello was the descendant of one of these Persian horses. Historians have claimed that there is not a shred of documentary evidence linking Pyrrhus and his troops with Metaponto, despite the fact that the Epiriote had fought a great battle with the Romans in 280 BC at Heraklea just 20 km to the south. Now we may say that there is at least a biological link between Metaponto and these momentous events. In sum, it becomes apparent how many previously invisible details archaeozoologists can show archaeologists and historians who seek to understand the canvas of ancient history.

Working with Sándor was always a learning experience. What impressed the younger members of the team especially was his deep seriousness and enthusiasm for his work, both reflected in his disciplined habits. His work day began at 9:00 a.m. and he maintained the same erect position at his desk, almost—except for the occasional ice cream break—without interruption until 6:00 p.m., day after day. There was the story that he told about examining 30,000 bones in one collection just to be sure that there was not one pig bone among them. It rang true, coming from him. There was no small talk, no wasted time in the laboratory—the "Banca"—during the sacred hours of work *(Fig. XIII)*. But after six, around the dinner table and with a glass of wine, he was another man, a fount of stories reflecting a long experience of places—many far from the common experience of the group—and of people, often with a surprising humorous outcome. He had met and known many, yet in his congeniality the seriousness about science, intellectual honesty, and friends was only just below the surface, and always with warmth. I, for one, have met few whose company and companionship I have enjoyed as much.

Joseph Coleman Carter
Austin
April 21, 2008

II. Sándor Bökönyi and Archaeozoology

László Bartosiewicz

"…a man of many devices, who wandered full many ways…"[1]

Sándor Bökönyi (1926–1994), whose studies form the core of this book, was one of the founders of modern archaeozoology during its world-wide emergence after World War II. This volume is both a tribute to and a continuation of his work, a token of appreciation of his persona as well as his scholarly achievements. As his former student I owe him a great debt; therefore it is difficult—and perhaps not even imperative—to be objective in a short introduction. This concise but informal biography may provide this book with a personal perspective.

The interests of an archaeozoologist are defined along the dimensions of chronology, geography, and animal species. These three vary spontaneously in one's work, but their combination is inevitably influenced by circumstances. A scholar's affinity to a subject is shaped by tutors, personal/professional contacts within the research community, and physical access to animal bone assemblages. The interplay between these factors may be best understood against the backdrop of broader cultural, political, and even economic events that limit or aid the realization of one's raw talent as well as conscious scholarly aspirations.

The list of publications by Sándor Bökönyi (see References), member of the prestigious Hungarian Academy of Sciences, shows that 10 percent of this rich scholarly heritage—21 out of 204—was conceived in cooperation with colleagues in Italy. His analysis of animal remains from the chora of Metaponto thus forms part of a long tradition. But how did a Hungarian researcher find his way to the shore of the Ionian Sea at the southern tip of the Apennine Peninsula?

First Came the Horses…

He was born in the village of Vállaj on the edge of the Ecsed Marshland in the Great Hungarian Plain, a settlement repeatedly forced into a marginal position during its history. In the 16th–17th century it lay on the frontier of the Ottoman Empire, while the 1920 Trianon Peace Treaty placed it on the present-day border with Romania. Vállaj lay near the famous "puszta," a rather desolate plain that has relied on extensive animal keeping for centuries. Environmental and historical factors thus both contributed to the perception of these rural backwaters as the "Wild East," perpetuated since the 19th century by romantic writers. Sándor, the talented son of the local school teacher, went from this traditional milieu to the high school of the Catholic Piarist order in the regional center, Debrecen. It is perhaps not surprising that, following graduation in 1944, he attended the Veterinary Faculty of the Palatine Joseph University in Budapest.

He never pursued a veterinary practice, however. In 1949, while still a student, he started working as a volunteer in the Museum of Natural History and the National Center for Museums and Monuments in Budapest, where he encountered a series of horse skulls and skeletons in storage belonging to the Migration Period (aka "Barbarian Invasion") ca. AD 300 to 700. These spectacular remains, with strong romantic and historical connotations, tended to be collected even before attention was paid to "ordinary" animal bones during excavations. The driving force behind the development of veterinary medicine since antiquity has been curing horses.[2] Horse finds, therefore, were most directly relevant to his skills as a graduate student, offering him an opportunity to put his veterinary expertise to the service of archaeology.

In reading about the horses, mules, and hinnies of the chora of Metaponto, the ancient compilation Mulomedicina by Vegetius Renatus (circa the 4th century AD) inevitably springs to mind. This synthesis of Greek veterinary treatises for Roman use offered a glimpse beyond Italy. Vegetius discussed horse keeping by "barbarians," tackling issues of intercultural influence resulting from colonial contacts with non-Roman, indigeneous horsemanship. What could be more relevant to the horses of Magna Grecia than the find of a complete horse skeleton in Tomb 316 at the Pantanello necropolis (Carter 1998b, 135) or

[1] Homer, Odyssey 1.1

[2] This is evident in scientific treatises, such as Xenophon's Art of Horsemanship, Aristotle's *Historia Animalium*, the *De Agricultura* of Cato, and Varro's *De Re Rustica III*.

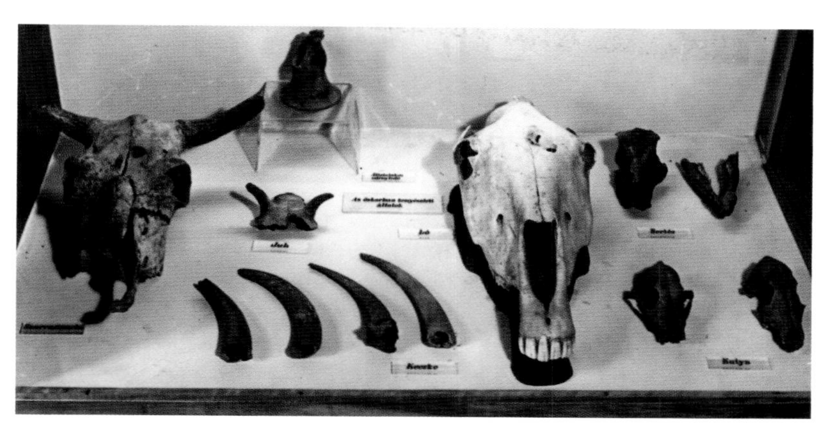

Figure XIII Archaeozoological finds from an exhibition in Hungary, early 1960s.

the study of variability in Roman period horses from the Pantanello Kiln Deposit? His chapters reveal that Sándor relied extensively on his early, hands-on experience with Scythian horse remains (Bökönyi 1952; 1954) in evaluating these animals.

Diffusion and Local Domestication

Because his early research had proven important, Sándor was granted a position at the Department of Archaeology of the Hungarian National Museum in 1951, where he established the Archaeozoological Collections *(Fig. XIV)*. The chronological scope of his research broadened to include an ever increasing number of prehistoric sites. This was not long after the realization by local archaeologists that food production must have arrived to the Carpathian Basin at the time of the Early Neolithic Körös culture. The bones of sheep and goat, not native to Europe, offered indubitable zoological evidence to support this hypothesis. Although the detailed roles played by diffusion, cultural transmission, and acculturation in the emergence of the Early Neolithic in Hungary are still being scrutinized (see Whittle 2007), Sándor contributed a key element to the discussion at its onset. As a result of his internationally acknowledged involvement in domestication research, he was invited to the Archaeological Institute of the Hungarian Academy of Sciences in 1973 to join a think-tank of prominent prehistorians of the time.

The area of modern-day Hungary, the Carpathian Basin, has served as an important geopolitical corridor between East and West since the early Neolithic. The resulting tumultuous history has made theories of diffusion and even invasion tangible and popular among Hungarian archaeologists. Having recognized mobility as represented in the bones of sheep and goat, Sándor dialectically argued for local domestication, especially in the case of cattle and domestic pig, the wild forms of which lived in the area. This is a recurrent theme in the evaluation of the Metaponto finds as well. Recently, however, there has been an effort to distinguish between the origins of these two species: while cross-breeding between wild and domestic pig is well known from the ethnographic record, large scale local domestication of the fierce aurochs seems to have been less likely than previously thought.

Window on the World

During the first third of his career, publications by Sándor dealt with archaeological sites in Hungary, reflecting only sporadic contacts in countries within the "East-Block" (Bökönyi and Kubasiewicz 1961; Bökönyi 1962a). With the slow political thawing in the wake of the Cold War in Hungary, his first work in Austria (Bökönyi 1965) was published within a decade after the 1956 popular uprising. His international activities, however, were really launched by a Ford Fellowship to the US in 1966–1967, followed by

Figure XIV Sándor Bökönyi in Oman, Jordan, 1976.

a six month study trip to Iraq in 1969, a year after the Arab Socialist Baath Party came to power and relations with Hungary improved. Studying important assemblages from southwest Asia during these visits spurred his interest in the region *(Fig. XV)*.

In the Debrecen high school, Latin and German were mandatory studies. German was the foreign language of choice in the pre-war school system and until recently remained of fundamental importance across Central Europe. This was highly fortunate for Sándor, because classical 19th century archaeozoology had also emerged in countries where German was the lingua franca of scholarly communication, including Denmark and Switzerland (Forchhammer et al. 1851; Rütimeyer 1861). A 1966 trip to the US also proved important for his career as it not only offered Sándor a new direction for research but also added the command of English to his foreign language skills. Equipped to read, write, and negotiate in both German and English, he was able to link two separate linguistic spheres in the academic world.

During the 1970s, the Great Hungarian Plain—the westernmost steppe region in Europe and the northwestern border zone of the distribution of prehistoric tell settlements—attracted increasing world interest, putting Sándor much in demand. His research linked the traditional "Germanic" school of archaeozoology to the Anglophonic prehistoric investigations in the Middle East that had started to rely more heavily on the use of natural sciences (Braidwood 1957). Anglo-American archaeological projects in the former Yugoslavia were another intellectual market for his professional skills. Eventually, his systematic investigations into the emergence of animal husbandry covered an entire region between Iran (Bökönyi and Bartosiewicz 2000), relevant areas in the Near East, the Balkans, the Apennine Peninsula, and ultimately France (Bökönyi and Bartosiewicz 1999).

The Long Way to Italy

These creative years, spanning the last two thirds of Sándor's career, yielded over 80% of his publications. Faunal reports from Italy began appearing in 1982. These research opportunities, however, largely originated in southwest Asia, where Sándor was employed as faunal analyst at Italian projects from Baluchistan to Anatolia and the Arab Peninsula, and where he met Lorenzo Costantini, already involved as a paleobotanist on the Metaponto project. The 1979 Iranian Revolution was of direct relevance to his career: although it forced Western expeditions out of that country, members maintained working contacts, which accounts for some of Sándor's research in Italy. He also taught as a visiting professor in Rome during 1983. This, with Costantini's introduction, led him to Metaponto.

The years highlighted by publications related to Italy *(Fig. XVI)* coincide with an important period in Sándor's career. In 1981 he was appointed director to the Archaeological Institute of the Hungarian Acad-

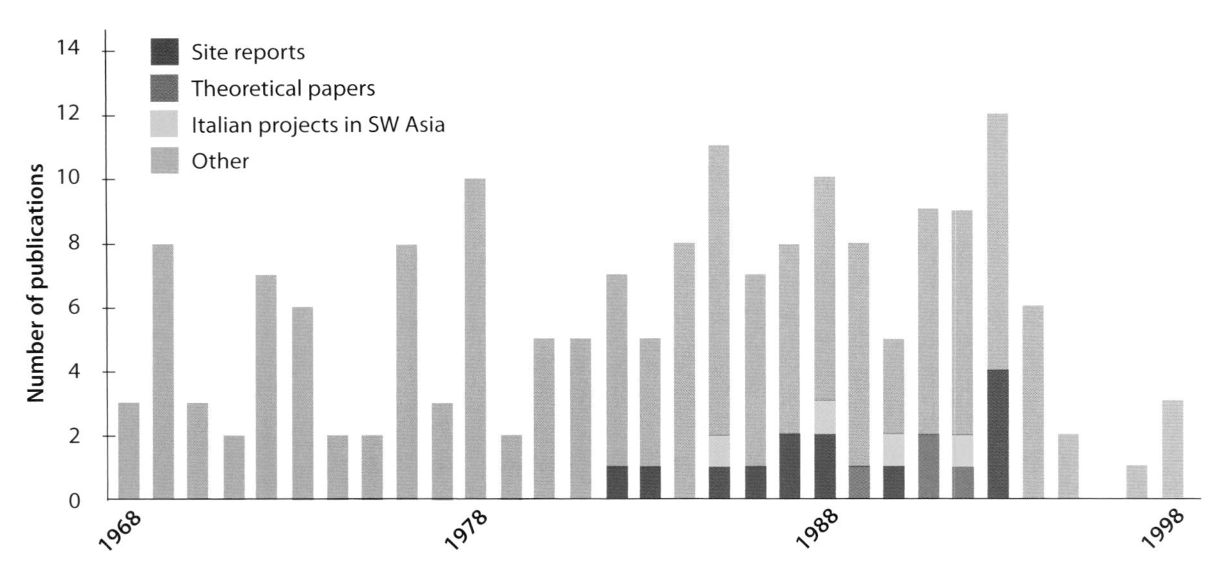

Figure XV The diachronic distribution of Sándor Bökönyi's publications relevant to Italian archaeology.

Figure XVI Sándor Bökönyi's former office in the Archaeological Institute, Budapest.

emy of Sciences and served 3 consecutive four year terms, supported by a democratic re-election in 1989 *(Fig. XVII)*. Owing to his scientific training and research experience in the United States,[3] he was seen as a representative of "New Archaeology." This intellectual trend could be seamlessly combined with the Marxist/historical approach heralded by V. G. Childe which was influential in Hungary.

The appointment was also an acknowledgment of Sándor Bökönyi's international stature that included working contacts in southwest Asia with his colleagues from Italy. One of the ways he invested these valuable connections was that he helped set up the first-ever Hungarian archaeological expedition on Italian soil. Excavations began in 1983 at the late 1st–3rd century AD Imperial period villa in San Potito di Ovindoli (Bökönyi 1986). This cooperative project has run successfully for a quarter of a century.

Where is Metaponto?

Naturally, this question does not concern the geographical location, but the academic position of this unique set of archaeological sites in Sándor Bökönyi's research. The answer is simple: it was the most complex assemblage he studied in Italy, rivalled in size only by the large Bronze Age settlement of Coppa Nevigata (Bökönyi and Siracusano 1987).

As an introduction to this volume, the size and zoological diversity of these sites is worth reviewing. When the number of animal taxa[4] is plotted against the number of bone finds, basic characteristics of the material from the chora of Metaponto may be recognized *(Fig. XVIII)*. They form a set of smaller and a set of larger assemblages respectively. The taxonomic richness—that is, the number of animals recognized—tends to increase with assemblage size. This trend is shown by a degressive curve in the graph drawn on the basis of all assemblages in the figure.[5] Assemblages above this line are rich in species relative to their sizes. Those below are less diverse than would be expected on the basis of their sizes.

While the faunal samples from smaller sites (with fewer than 1000 identifiable bones) largely fall onto the trend line, the unusually rich, large sites from Pantanello (especially from the 7th–3rd c. BC Greek sanctuary) contained bones from a whole range of curious wild animals. The opposite holds true for the 7th–6th century BC settlement of Incoronata, where the large assemblage is characterized by a monotonous repertoire of a few animal species.

The sizes and richness of these faunal assemblages convincingly demonstrate that the work Sándor invested in this material, in a way, crowned his research in Italy. It was especially fortunate that he found reliable research partners and good company in the team headed by Joseph Coleman Carter, who also became an appreciated friend.

What is very special about this volume is that it is entirely devoted to animal remains from a complex range of sites in one geographical area. Sándor addressed Italian archaeologists in the Museo Pigorini of Rome on the 30th of May, 1983 *(Fig. XIX)*. Referring to Forchhammer's Danish research team of a prehistorian, a geologist, and a zoologist who inadvertently founded archaeozoology in 1851, he exclaimed:

[3] His numerous guest professorships included two semesters of teaching at the University of California Los Angeles in 1970-1971, at a time when "New Archaeology" came to fruition in the United States.

[4] Although the remains of different animal classes, such as mammals, birds, and fish, behave differently in such comparisons (Bartosiewicz and Gál 2007), they were pooled for the sake of simplicity in Figure XVIII.

[5] Mathematically, this can be described with the exponential equation: $y = 2.702 \times 0.248$ ($r = 0.796$, $df=14$, $P \leq 0.01$). These parameters are indicative of a close relationship between the size and diversity of these assemblages.

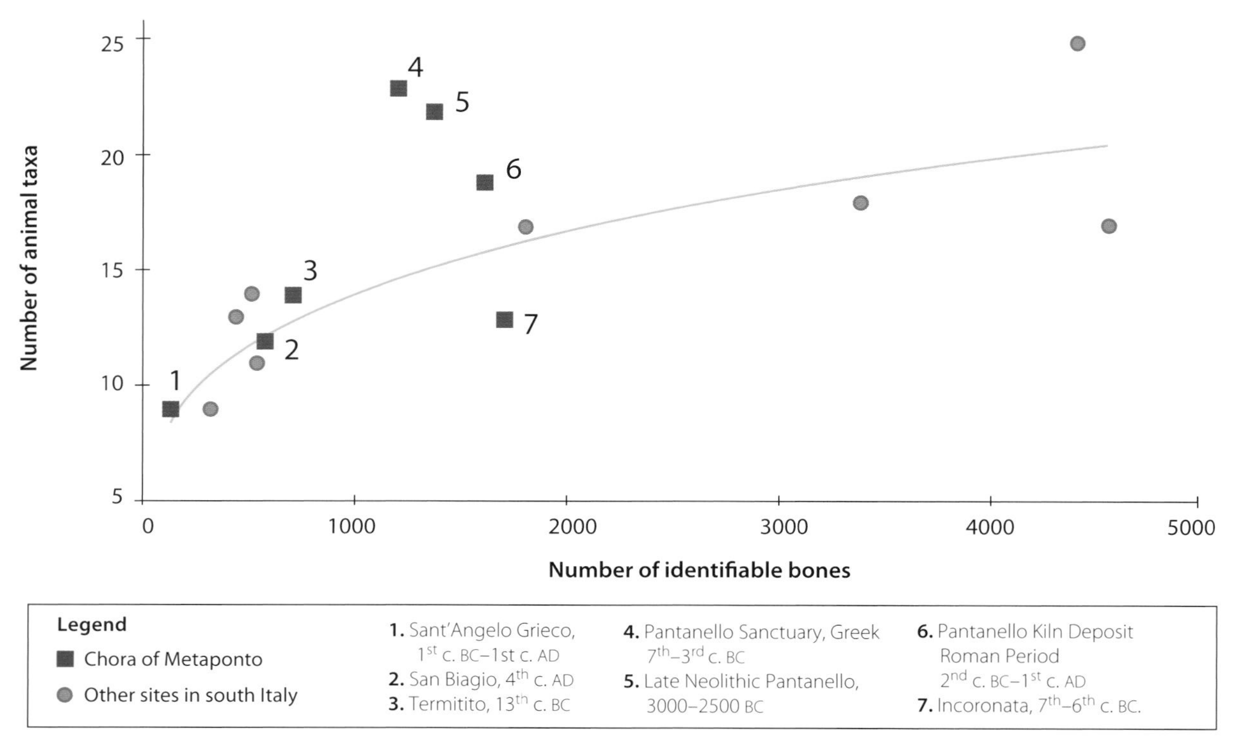

Figure XVII The size and taxonomic richness of faunal assemblages from South Italy published by Sándor Bökönyi.

"I just cannot believe that 140 years later the necessity of such cooperations should be stressed and stressed again" (Bökönyi 1992, 400).

A quarter of a century later, almost 160 years after that famous alliance, the director of a major archaeological project at Metaponto devoted resources, both material and intellectual, to the very same type of multidisciplinary inquiry. The posthumous publication of this work is likewise impressive. Rather than abandoning the manuscript after so many years, as often happens after someone passes away, Professor Carter involved a new generation of Hungarian archaeozoologists. They have digitalized, updated, and enhanced the research with special aspects of archaeozoology (taphonomy, bone manufacturing) that were less developed at the time Sándor worked at the site. The materials were illustrated by a trained bone expert, as he would have wished.

Figure XVIII Sándor Bökönyi lecturing in Rome, May 1983.

This book stands for continuity: not only the spectacular continuity of ancient civilizations along the picturesque shores lapped by the Ionian Sea, where the sight of grazing flocks, scent of strawberries, and the night call of little owls in pine trees made Sándor feel very much at home; but also the intrinsic continuity of scholarly thought and cooperation that has linked three generations of archaeozoologists in this pioneering study of the ancient countryside.

László Bartosiewicz
April, 2008

1

Animal Husbandry from the Late Neolithic through the Roman Period

Sándor Bökönyi

Edited by László Bartosiewicz with Erika Gál

Editor's Note: This is an updated but unabridged version of the manuscript written by the late Sándor Bökönyi (1926–1994). It was singled out by the author for special analysis as a deposit perfectly suited for reconstructing Roman Period animals. Publication of his archaeozoological work for the Metaponto Project involved the cooperation of several animal bone experts. Éva Nyerges entered Bökönyi's large body of hand-written data, including all the identifications and notes for over 10,000 bones, into Excel spreadsheets during 2004–2005. This task involved the standardization and translation of complex technical terms, often abbreviated, written in an admixture of Bökönyi's Latin, English, and Hungarian.

These basic data were verified and re-analyzed on location by Erika Gál during 2007. She calculated new numerical tables in accordance with the final phasing of all sites. This tedious work resulted in only minor shifts in Bökönyi's results, but presented a hurdle that had to be addressed professionally. She also verified and arranged bone measurements in the Appendix according to the protocol established by Angela von den Driesch (1976), and took photographs for this publication.

Minimal editorial changes were made to the text itself. New information (listed separately in footnotes after the original references) was added by László Bartosiewicz and Joseph Carter on the basis of recent research. Bökönyi set aside a number of specimens for drawing. This artwork was completed by Anna Biller, another trained archaeozoologist, in the summer of 2007, with additional renderings made in 2008.

LB

Introduction

The study of ancient domestic and wild animals through their remains is essential for understanding the development of animal husbandry and hunting as human occupations in a locale. The remains reveal an important sector of the local economy: the acquisition of animal protein and fat, as well as hides, horns, antlers, sinews, and other raw materials of animal origin. When animal bone assemblages span a large chronological sequence, they can provide exceptional insight into the dynamics of cultural development.

Metaponto and its environs represent an ideal subject for such studies. First, the area consists of a uniform geographical unit, where there are only minor differences in terms of the longer or shorter distance to the river or sea, as well as minimal changes in elevations. Second, the climate was likewise uniform, so that relief, soil, climate, and vegetation generally formed one large single habitat; this, however, does not exclude the existence of small microhabitats (e.g., the bush strips along the rivers or certain deeper valleys). Third, archaeological sites (mainly settlements) of different types have been excavated from periods between the Neolithic and the end of the Roman Empire. With such a seemingly unbroken faunal series, one can study the development of both the domestic and wild fauna against the backdrop of a largely standardized environment, without the disturbing effects of habitat differences. Indubitably, possible climatic change may have affected the fauna, but still they can be more easily detected here than in faunal assemblages originating from different environments.

The animal bone samples of the Metaponto area offer a good opportunity to address at least several questions regarding changes in wild fauna and domesticates. First of all, one can get an idea about the remnants of the original wild fauna that had flourished there before animal husbandry first arrived from the

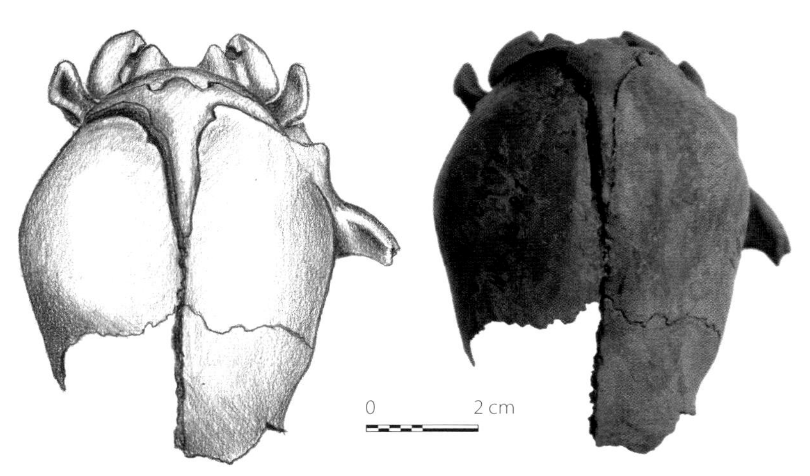

Figure 1.1 Frontal aspect of a young dog's neurocranium, Pantanello Sanctuary (PZ-78-455 B).

East, sometime around the beginning of the 5th millennium BC. Thereafter, one may follow the changes of the wild fauna caused by the gradually increasing human population and steadily growing number of domestic animals. In this respect, even the extinction or disappearance of certain wild species may be observed. Still, the relatively small number of wild animal bones seems insufficient for the demonstration of climatic changes which can be so excellently outlined in Western, Central, and Eastern Europe.

There are also questions regarding the acclimatization of the earliest domesticates, particularly those which had no local wild forms and were entirely alien to the Apennine Peninsula. The Caprinae subfamily (Gray, 1821), including sheep and goat, is a good example. Their introduction in Italy was closely connected to what happened in the northern part of the Balkans in Central and Eastern Europe, where, eventually, the number of caprines strongly diminished owing to the lack of domesticable wild forms. At least in principle, stocks of cattle and pig may have grown through the local domestication of aurochs and wild swine that were certainly present, if not exactly in such large numbers as in the aforementioned areas (Bökönyi 1974, 103; 1977-82, 346; 1983a, 242; 1985, 194; 1985, 184).[1] The time of earliest arrival of new domestic species, e.g., horse, ass, and domestic hen, is also important, as well as the first occurrence of conscious and selective animal breeding, which led to new, well-determinable breeds. Another significant question is the improvement of feeding, both in quantitative and qualitative respects, which also influenced the general increase of the average withers height.

Such a diachronic study requires a complex range of analyses. Description of the domestic animal species and/or breeds encountered is indispensable, along with estimating the size, body proportions, horn size/form, hornlessness, and other marks of fresh or early domestication. Other aspects to consider are the types of exploitation for the different domestic species, their changes between different periods, as well as pathological cases and their causes. Butchering techniques applied to large and small animal species also requires systematic recording; marks of gnawing by dogs and other scavengers are also of interest. While B. Cabaniss (1983) and S. Scali (1983) published preliminary reports of some parts of the Metaponto faunal remains, they were only able to touch upon a limited number of the problems that can be discussed in connection with this valuable animal bone assemblage. The purpose of this volume is to approach the subject from a multidisciplinary point of view and so provide a multifaceted and more comprehensive picture of the material.

The Materials

The animal remains found in the sites of the Metaponto area can be divided into three different groups. The first group, constituting the overwhelming majority, is represented by kitchen refuse. This means that

	Cattle	Sheep	Goat	Horse	Dog	Chamois	Fox	Wolf	TOTAL
Scapula	-	-	-	-	-	-	1	-	**1**
Humerus	-	-	-	-	-	-	1	-	**1**
Radius	7	2	-	2	-	-	1	-	**12**
Metacarpus	12	2	-	7	-	1	-	-	**22**
Femur	-	1	-	-	-	-	-	-	**1**
Tibia	-	-	-	2	1	-	-	1	**4**
Metatarsus	12	3	1	6	-	-	-	-	**22**
TOTAL	**31**	**8**	**1**	**17**	**1**	**1**	**3**	**1**	**63**

Table 1.1 Complete main extremity bones, Chora of Metaponto.

[1] Recent DNA evidence has fine-tuned what looked like a clear-cut effect of local domestication at the time this manuscript was written. See note 2 for details.

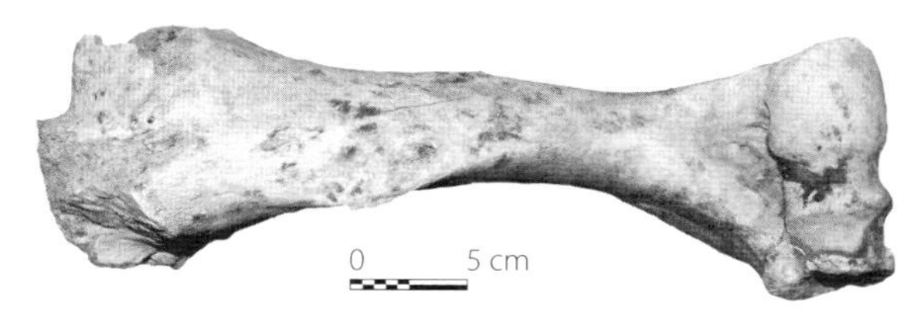

Figure 1.2 Complete cattle humerus, Pantanello Kiln Deposit (PZ 81-462B-8).

the remains are in a poor state of preservation, with no complete skeletons or large, articulated skeletal parts with bones in anatomical order. Complete skulls are also missing, and only one nearly intact neurocranium of a young dog, dated to the 5th century BC, was found at Pantanello site *(Fig. 1.1)*. The number of horn cores and horn core fragments is also very small, and complete long bones represent only an insignificant majority of this part of the sample. (See Table 1.1, in which not even all the bones belong to this group; some of them are from animal sacrifices, others from modern carrion, buried into the soil of the site.) All other bones were broken up for the marrow or simply cut into small pieces, since otherwise they could not have been put into cooking pots.

The data from Table 1.1 show that almost half of the complete bones (28 specimens, thus 44.4%) come from the Roman period Pantanello Kiln Deposit (the small and numerous metapodial bones from pigs and carnivores were not included in this table). The entire Greek and Roman site of Pantanello yielded 17 complete bones (27.0%), while the other sites were poorer in this regard. One site, San Biagio, yielded no complete bones at all.

Of the total number of complete bones, most came from cattle *(Fig. 1.2)*, with 31 examples (49.2%). Horse ranked second with 17 bones (27.0%), and sheep *(Fig. 1.3)* were still worth mentioning with 8 bones (12.7%). All other species lagged behind. Among the remains of the most typical domestic "meat" animal, the pig, there were conspicuously no complete long bones whatsoever. This might have two explanations: first, most pigs were killed at an immature age when the epiphyseal ends of the long bones had not

Figure 1.3 Complete sheep metacarpal, Pantanello Kiln Deposit (PZ 81-667B-1).

yet completely fused; second, the pig bones were the most badly fragmented. Complete ass and dog bones were also missing. The rarity of the first species and the generally poor preservation of bones from the latter make this understandable. It was also evident that among the remains of the main meat-providing wild species no fully preserved long bones survived, in spite of the sometimes high frequency of wild animal remains.

Among the bones of the different extremity segments, those of the relatively meatless parts, the metapodia, were the most frequent, with 22 metacarpals

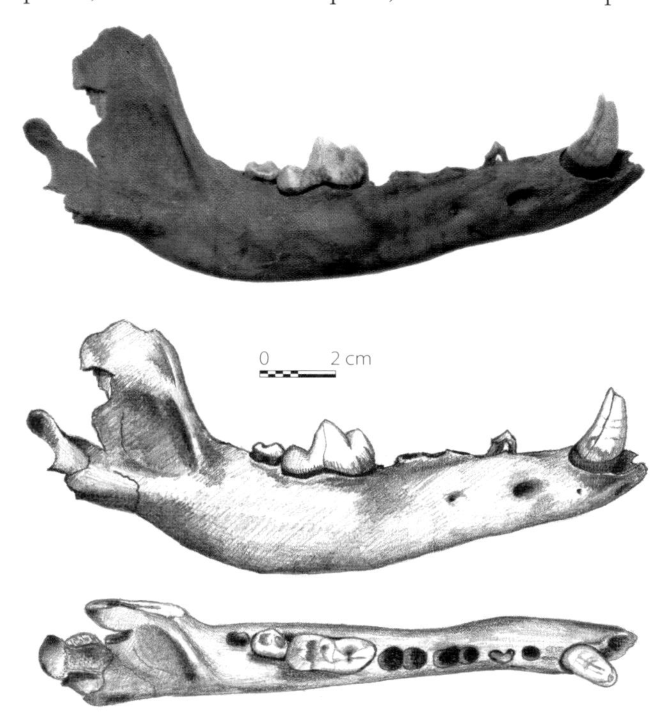

Figure 1.4 Buccal and occlusal aspects of a right wolf mandible found in a ritual context, Pantanello Sanctuary (PZ 78 359 B).

and 22 metatarsals making up 69.8% of the sample (the ratio from the fore and hind limbs was the same, and as noted before, the count does not include the smaller metapodial bones from pigs and carnivores). The explanation for this frequency may be that these sturdy bones contain very little marrow and are hard to break up. Moreover, they could have been left in place to provide support during the skinning of the animal. The second most commonly preserved complete bone was the radius, represented by 12 specimens (19.0%). All other skeletal parts were infrequent.

The second group of bones comes from animal sacrifices. This includes the right mandible of a wolf

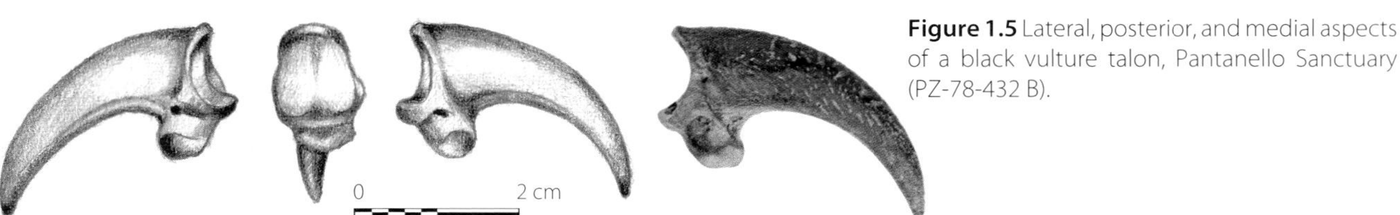

Figure 1.5 Lateral, posterior, and medial aspects of a black vulture talon, Pantanello Sanctuary (PZ-78-432 B).

(Fig. 1.4) found at the Pantanello Sanctuary (6th–3rd century BC) or possibly the bones of a black vulture discovered in a later 4th–3rd century BC deposit at the same site *(Fig. 1.5)*, as S. Scali suggests in his preliminary report on a portion of the Metaponto animal bone assemblage (Scali 1983).

The third group of archaeozoological finds is represented by more complete skeletons: a horse and a wolf skeleton found in the Pantanello Necropoleis (6th–3rd century BC); an incomplete skeleton of a brown hare (the contemporaneity of which, rather than intrusive nature, is shown by transversal cut marks on its collum scapulae) from Late Neolithic Pantanello; and a nearly complete goat skeleton from the Late Roman farmsite at San Biagio (late 3rd–4th century AD). The horse burial yielded a fragmented skull, both mandible halves, and a practically complete skeleton; the wolf remains consisted of only a cranial fragment and an incomplete skeleton with extremely fragmented bones; the hare skeleton was missing the skull and a mandible. Both animals died at mature ages and were seemingly interred at the settlement. They are certainly contemporaneous with the other burials at the Necropoleis, as is demonstrated by the color of the bones and their state of preservation.

The Wild and Domestic Fauna and Their Development

The sites in the Metaponto area, which span roughly three millenia from the Late Neolithic to the Late Roman period (4th century AD), yielded a rich bone assemblage with no less than 21 animal species, including 15 mammals, 3 birds, 1 tortoise, and 1 fish *(Table 1.2)*. This is a major number of wild animal species, particularly as regards mammals. The 16 mammalian taxa in this group correspond to 70% of the wild mammal species generally found during the time of the Mesolithic–Neolithic transition in Europe (Jarman 1972, Fig, 1), and equal the number of the wild animal species identified at six early Neolithic sites from Southern Italy (Bökönyi 1985, Table 1). Lynx, marten, bear, otter, and beaver are the only animals missing among the commonly occurring wild species. These forest-dwelling carnivores, however, tend to be very wary of large, urban settlements and are rarely hunted for food.

Most of the wild mammals were found at Late Neolithic Pantanello, despite the fact that, in terms of the number of bone specimens, wild animals made up only 6.17% of the sample *(Table 1.3)*. Only fallow deer,

Species	Latin name
Aurochs	*Bos primigenius* (Boj., 1827)
Ibex	*Capra ibex* (L., 1758)
Chamois	*Rupicapra rupicapra* (L., 1758)
Red deer	*Cervus elaphus* (L., 1758)
Fallow deer	*Dama dama* (L., 1758)
Roe deer	*Capreolus capreolus* (L., 1758)
Wild swine	*Sus scrofa* (L., 1758)
Wild ass	*Equus hydruntinus* (Reg., 1907)
Wild cat	*Felis silvestris* (Schreb., 1777)
Weasel	*Mustela nivalis* (L., 1758)
Badger	*Meles meles* (L., 1758)
Fox	*Vulpes vulpes* (L., 1758)
Wolf	*Canis lupus* (L., 1758)
Brown hare	*Lepus europaeus* (Pall., 1778)
Vole	*Arvicola* sp.
Small rodent	Rodentia
Wild duck	*Anas* sp.
Black vulture	*Äegypius monachus* (L., 1758)
Wild birds	Aves
Greek tortoise	*Testudo hermanni* (L., 1758)
Fish	Pisces sp.

Table 1.2 List of wild animal species, Chora of Metaponto.

Species List	Subtotals	Specimens	Percent
Cattle *Bos taurus* L.		147	11.52
Sheep *Ovis aries* L.	148		
Goat *Capra hircus* L.	34	1,038	81.34
Sheep/Goat *Ovis/Capra*	856		
Pig *Sus scrofa dom.* L.		88	6.90
Dog *Canis familiaris* L.		3	0.24
Domestic animals subtotal		**1,276**	**100.00**
Aurochs *Bos primigenius* Boj.	3		
Ibex *Capra ibex* L.	2		
Chamois *Rupicapra rupicapra* L.	3		
Red deer *Cervus elaphus* L.	2	19	22.62
Roe deer *Capreolus capreolus* L.	1		
Wild swine *Sus scrofa fer.* L.	7		
Wild ass *Equus hydruntinus* Reg.	1		
Weasel *Mustela nivalis* L.	1	3	3.57
Fox *Vulpes vulpes* L.	2		
Brown hare *Lepus europaeus* Pall.		34	40.48
Mallard *Anas platyrhynchos* L.	5		
Partridge *Perdix perdix* L.	2	9	10.71
Rook *Corvus frugilegus* L.	1		
Birds Aves sp.	1		
Turtle Chelonia		18	21.43
Fish Piscis sp.		1	1.19
Wild animals subtotal		**84**	**100.00**
TOTAL		**1,360**	

Table 1.3 Species and counts from Pantanello, Late Neolithic.

wild cat, wolf, vole, small rodents, and (among birds) black vulture were missing at this site. This range of diversity tends to appear at many Neolithic sites.

The ratio of wild animal bones at Late Bronze Age Termitito was even higher (15.9%), although the actual number of wild mammal species was only six, thus essentially lower *(Table 1.4)*. In South Italy there is a Bronze Age site, Coppa Nevigata, where excavations produced a large animal bone assemblage (11,391 identifiable specimens). Here, both the ratio of the wild animal bones (7.98% to 22.62%) and the number of wild mammal species (4 to 13) increased

Species list	Subtotals	Specimens	Per cent
Cattle *Bos taurus* L.		148	25.43
Sheep *Ovis aries* L.	47		
Goat *Capra hircus* L.	2	327	56.19
Sheep/Goat *Ovis/Capra*	278		
Pig *Sus scrofa dom.* L.		97	16.67
Ass *Equus asinus* L.		1	0.17
Dog *Canis familiaris* L.		9	1.54
Domestic animals subtotal		**582**	**100.00**
Ibex *Capra ibex* L.	1		
Red deer *Cervus elaphus* L.	31	37	33.65
Roe deer *Capreolus capreolus* L.	1		
Wild swine *Sus scrofa fer.* L.	4		
Wild cat *Felis silvestris* Schreb.		1	0.90
Brown hare *Lepus europaeus* Pall.		1	0.90
Birds Aves sp.		1	0.90
Turtle Chelonia		70	63.65
Wild animals subtotal		**110**	**100.00**
TOTAL		**692**	

Table 1.4 Species and counts from Termitito, Late Bronze Age.

Species list	Subtotals	Specimens	Percent
Cattle *Bos taurus* L.		381	24.00
Sheep *Ovis aries* L.	47	664	41.81
Goat *Capra hircus* L.	19		
Sheep/Goat *Ovis/Capra*			
Pig *Sus scrofa dom.* L.		518	32.62
Horse *Equus caballus* L.		10	0.63
Dog *Canis familiaris* L.		14	0.88
Hen *Gallus domesticus* L.		1	0.06
Domestic animals subtotal		**1,588**	**100.00**
Ibex *Capra ibex* L.	2	62	77.50
Red deer *Cervus elaphus* L.	53		
Roe deer *Capreolus capreolus* L.	1		
Wild swine *Sus scrofa fer.* L.	6		
Brown hare *Lepus europaeus* Pall.		2	2.50
Turtle Chelonia		16	20.00
Wild animals subtotal		**80**	**100.00**
TOTAL		**1,668**	

Table 1.5 Species and counts from Incoronata, 8th–6th century BC.

between the Early and Late Bronze Age (Bökönyi and Siracusano 1987, 205f). In North Italy the ratios of wild animals and the numbers of wild mammal species were lower in Bronze Age sites (Riedel 1976, 9; 1977, 69; 1979, 44; 1981, 124), though this could be, at least in some cases, related to the small number of the identified animal remains.

At Incoronata (8th–6th century BC), where the importance of hunting strongly decreased, as shown by the 4.8% ratio of the wild animals, the number of wild species was six, all "meat" animals, including five mammal species and a tortoise *(Table 1.5)*.

In addition to the remains of domestic animals recovered from the drainage channel (cloaca) at the Metaponto Sanctuary (4th century BC), a large bovine bone, probably from aurochs, was the only specimen possibly representing a wild species *(Table 1.6)*. This assemblage, however, is too small to allow far-reaching conclusions, so no percentages were calculated.

Species list	Specimens
Cattle *Bos taurus* L.	56
Goat *Capra hircus* L.	1
Horse *Equus caballus* L.	2
Domestic animals subtotal	**59**
Aurochs (?) *Bos primigenius Boj.*	1
Wild animals subtotal	**1**
TOTAL	**60**

Table 1.6 Species and counts from the drainage channel (cloaca) Metaponto Sanctuary, 4th century BC.

The Greek and Roman periods from the complex at Pantanello (6th century BC–3rd century AD) yielded the highest number of wild animal species—altogether 13—all of which were mammals *(Table 1.7)*. In comparison to the entire assemblage from the Metaponto area, only five animals were missing: chamois, wild ass, wild cat, weasel, and brown hare. The assemblage in which this relatively high number of wild mammal species occurred was not very large, although it had a comparatively high ratio of wild animal bones. Both the large number of wild mammal species and the high ratio of wild animals were surprising, because one would have expected less reliance on hunting in a "civilized" Greek colonial town and later in a well populated area of the Roman Republic and Empire.

Additional remains from wild animals can be found after the Roman conquest in the 3rd century BC, when the town itself "was reduced to the small area of a castrum or praesidium" and the number of inhabited places decreased markedly (Carter 1983a, 39; D'Annibale 1983, 11). The small animal bone sample from the Pantanello Necropoleis *(Table 1.8)* includes the complete skeleton of a wolf, found in Tomb 321 (Carter 1983b, 15).

No wild animals bones were found among the handful of ordinary, domestic ungulate remains in the Greek Pit at Pantanello *(Table 1.9)*. This feature, with

Species list	Subtotals	Specimens	Percent
Cattle *Bos taurus* L.		553	52.74
Sheep *Ovis aries* L.	20	227	22.06
Goat *Capra hircus* L.	5		
Sheep/Goat *Ovis/Capra*	202		
Pig *Sus scrofa dom.* L.		58	5.63
Horse *Equus caballus* L.		148	14.38
Ass *Equus asinus* L.	5	8	0.78
Ass/mule *Equus asinus/ Equus asinus x Equus caballus*	3		
Dog *Canis familiaris* L.		32	3.01
Cat *Felis catus* L.		1	0.10
Hen *Gallus domesticus* L.		2	0.19
Domestic animals subtotal		**1,029**	**100.00**
Aurochs *Box primigenius* Boj.		11	7.00
Ibex *Capra ibex* L.		1	0.64
Red deer *Cervus elaphus* L.		91	57.96
Fallow deer *Dama dama* L.		4	2.54
Roe deer *Capreolus capreolus* L.		2	1.27
Wild swine *Sus scrofa fer.* L.		5	3.18
Badger *Meles meles* L.		1	0.64
Fox *Vulpes vulpes* L.		5	3.18
Wolf Canis *lupus* L.		6	3.82
Brown hare *Lepus europaeus* Pall.		2	1.27
Black vulture *Aegypius monachus* L.		8	5.09
Rodent Rodentia		2	1.27
Turtle Chelonia		19	12.10
Wild animals subtotal		**157**	**100.00**
TOTAL		**1,186**	

Table 1.7 Species and counts from the Pantanello Sanctuary, 6th–3rd century BC.

Species list	Specimen	Tomb number
Cattle *Bos taurus* L.	11	37, 127, 128, 301, 315
Sheep *Ovis aries* L.	22	128, 263, 264, 307
Goat *Capra hircus* L.	14	128, 263, 264
Sheep / Goat *Ovis / Capra*	67	114, 127, 128, 263, 264, 307, 347
Pig *Sus scrofa dom.* L.	3	347
Horse *Equus caballus* L.	150	316
Mule *Equus asinus x Equus caballus*	41	62
Domestic Animals Sub.	**308**	
Wolf *Canis lupus* L.	90	321
Wild animals subtotal	**90**	
TOTAL	**398**	

Table 1.8 Species and counts from the Pantanello Necropoleis, 6th–3rd century BC.

Species list	Specimens
Cattle *Bos taurus* L.	3
Sheep / Goat *Ovis / Capra*	9
Pig *Sus scrofa fer. L.*	9
Ass *Equus asinus* L.	1
Domestic animals	**22**
TOTAL	**22**

Table 1.9 Species and counts from the Pantanello Greek Pit, 6th century BC–3rd century AD.

materials from different periods, had been dug into the Late Neolithic settlement layers.

The Pantanello Kiln Deposit, dated roughly from the 2nd century BC to the 1st century AD, contained the largest animal bone sample in the Metaponto area. With a total of 1,598 identifiable specimens, the ratio of identifiable wild animal remains was 11.26%, while the number of wild animal species amounted to ten *(Table 1.10)*. Nine of these were mammals (an over-

Species list	Subtotals	Specimens	Percent
Cattle *Bos taurus* L.		641	45.20
Sheep *Ovis aries* L.	41		
Goat *Capra hircus* L.	7	395	27.85
Sheep/Goat *Ovis/Capra*	347		
Pig *Sus scrofa dom.* L.		108	7.63
Horse *Equus caballus* L.		140	9.88
Ass *Equus.*	4	6	0.42
Ass/mule *Equus asinus/ Equus asinus x Equus caballus*	2		
Dog *Canis familiaris* L.		121	8.53
Hen *Gallus domesticus* L.		7	0.49
Domestic animals subtotal		**1,418**	**100.00**
Aurochs *Box primigenius* Boj.		11	6.11
Ibex *Capra ibex* L.		1	0.56
Red deer *Cervus elaphus* L.		114	63.33
Fallow deer *Dama dama* L.		9	5.00
Roe deer *Capreolus capreolus* L.		3	1.67
Wild swine *Sus scrofa fer.* L.		20	11.11
Badger *Meles meles* L.		2	1.11
Fox *Vulpes vulpes* L.		11	6.11
Brown hare *Lepus europaeus* Pall.		1	0.56
Tortoise *Testudo cf. hermanni* L.		8	4.44
Wild animals subtotal		**180**	**100.00**
TOTAL		1,598	

Table 1.10 Species and counts from the Pantanello Kiln Deposit, 2nd c. BC–1st c. AD.

whelming majority of which were red deer) and the tenth was the Greek tortoise.

In the small sample (125 specimens) from the Roman farmhouse at the site of Sant'Angelo Grieco (6th century BC–1st century AD) the remains of only three wild ungulates—two deer species and wild boar—were found *(Table 1.11)*.

The excavation of the Roman villa at San Biagio (late 3rd–4th century AD) yielded a medium-size bone assemblage of 570 specimens (*Table 1.12)*. Here there was a relatively low ratio (3.15%) of wild animals representing only four species (three mammals and the Greek tortoise).

All wild animal species—particularly the herbivores—have habitats that are determined by climate, topography, soil, water, vegetation, competing species, etc. The presence of certain animals supposes certain habitat types around the site or within a two-day walking distance, so that prey can be carried from the kill site to the settlement efficiently. Based on the composition of these species—mainly mammals—one can reconstruct hypothetically the existence of four habitat types in and around the Metaponto area between the Late Neolithic and the Late Roman Imperial periods:

1. Grassland with forested spots
2. Extensive, dense forest
3. Gallery forest along rivers or watercourses
4. Medium-range or high mountain areas

For animals such as the aurochs, wild ass, brown hare, and Greek tortoise, Type 1 is ideal. Fallow deer, roe deer, and wild cat are characteristic of the open forests in Type 1 and Type 3. Red deer prefer Type 2 large and dense forest. Although they can survive in the first and third types, only smaller forms develop in such open habitats. This does not seem to be the case in the Metaponto area even in Roman times, when antlers from large individuals could be found.

Wild swine need forest or shrub cover and therefore prefer dense or gallery forest. Badgers live in forest and open landscape with wooded spots. The fox can survive in any place with at least some forest, but avoids humid habitats such as gallery forests. Ibex and chamois originally lived in medium range mountains and have sometimes been described at sites in those locales; they were pushed into the high mountains only in early historic times by the expansion of human occupation. Not even medium-size mountains are to be found in the Metaponto area and its close vicinity, and hunters today would have to cover long distances to kill these game. Finally, weasel and wolf can live in any of the four habitat types (van den Brink 1957, 70ff). As for birds, wild ducks require water, while remote mountains and plains are the preferred habitat of the black vulture (Peterson, Mountfort, and Hollom 1966, 9f).

Turning our attention to domestic fauna, one can immediately observe that animal husbandry did not change in South Italy between the Early and Late Neolithic. In the Late Neolithic assemblage from Pantanello, exactly the same caprine-based domestic fauna can be found as in the early Neolithic samples from Rendina in Basilicata (Bökönyi 1977-82, Table 1), from the Tavolière Plain (Bökönyi 1983a, 239, Table 2) and from Passo di Corvo (Sorrentino 1983, Tab. 75). At Pantanello, caprines represented 81% of the domestic animal remains, while cattle bones were less than 12%, and pigs did not reach 7%. The contribution of dog remains was insignificant.

Among the domestic animal remains from Late Bronze Age Termitito, caprines still provided the absolute majority of domestic animal remains, though their frequency was much lower, only 56.19% compared to earlier Neolithic sites. Cattle ranked second, and pig third. Dog bones were rare, and the discovery of domestic ass was a great surprise. This find and the materials uncovered at the Late Bronze Age phase of Coppa Nevigata (Bökönyi and Siracusano 1987, 707) provide the earliest evidence for the domestic ass in Italy. In terms of the origins of these animals, there are several possibilities. They could have come directly from Egypt—where their wild forms lived and their first domestication took place—or from Greece, where they first appeared at the settlement of Kastanas in the Middle Bronze Age (Becker 1986, 72, Taf.30). Another source might be Asia Minor, where there were close trade connections and where the earliest domestic asses appeared as early as the Late Chalcolithic, that is, the later half of the 4th millennium BC (Bökönyi 1983b, 589).

Species list	Subtotals	Specimens
Cattle *Bos taurus* L.		61
Sheep *Ovis aries* L.	2	15
Sheep/Goat *Ovis/Capra*	13	
Pig *Sus scrofa dom.* L.		8
Horse *Equus caballus* L.		6
Dog *Canis familiaris* L.		2
Hen *Gallus domesticus* L.		1
Domestic animals subtotal		**93**
Red deer *Cervus elaphus* L.		28
Fallow deer *Dama dama* L.		1
Wild swine *Sus scrofa fer.* L.		3
Wild animals subtotal		**32**
TOTAL		**125**

Table 1.11 Species and counts from Sant'Angelo Grieco, 6th century BC–1st century AD (primarily 2nd c. BC–1st c. AD).

Species list	Subtotals	Specimens	Percent
Cattle *Bos taurus* L.		31	5.62
Sheep *Ovis aries* L.	6	323	58.51
Goat *Capra hircus* L.	194		
Sheep / Goat *Ovis / Capra*	123		
Pig *Sus scrofa dom.* L.		163	29.53
Horse *Equus caballus* L.		4	0.73
Dog *Canis familiaris* L.		2	0.36
Hen *Gallus domesticus* L.		29	5.25
Domestic animals subtotal		**552**	**100.00**
Red deer *Cervus elaphus* L.		3	16.67
Wild swine *Sus scrofa fer.* L.		2	11.11
Brown hare *Lepus europaeus* Pall.		3	16.67
Rodent Rodentia		2	11.11
Turtle Chelonia		8	44.44
Wild animals subtotal		**18**	**100.00**
TOTAL		**570**	

Table 1.12 Species and counts from San Biagio, late 3rd–4th c. AD.

At Incoronata (8th–6th century BC) there was an interesting change in the domestic fauna, namely that caprines lost their absolute majority, constituting only 41.77% of the domestic animal bones. They still, however, remained the leading group, closely followed by pig and then cattle. All other species were infrequent, but two new domestic animals appeared: horse and domestic hen. One would have expected to see domestic horses already appearing in the sample from Termitito, because the horse arrived in North Italy during the Early Bronze Age Polada culture (Azzaroli 1972, 1985, 129; Riedel 1977, 69) and reached South Italy by the Late Bronze Age (Bökönyi and Siracusano 1987, 206). It is possible that the absence of domestic horse bones at Termitito may be explained by the small sample size.

The presence of the domestic hen, represented by a tibiotarsus fragment from Incoronata, is the earliest occurrence in Italy to our most recent knowledge. A typical species of modern animal husbandry, this find solves Gandert's dilemma as to whether the domestic hen reached Italy through Greek colonization or through Etruscan trade with Asia Minor (Gandert 1952, 117; 1953, 74).

It is fruitless to examine the smaller assemblages from the Metaponto area—the Greek Pit at Pantanello, the cloaca materials from the Metaponto Sanctuary, the Pantanello Necropoleis, and Sant'Angelo Grieco—to learn more about the development of the domestic fauna. These sites did not produce samples of representative sizes that would reliably show changes between the ratios of different domestic animals. Still, these smaller assemblages must be considered in the development of individual domestic species.

A comparison of the Pantanello site assemblage as a whole (6th century BC–3rd century AD) to that of the Pantanello Kiln Deposit (2nd century BC–1st century AD) and San Biagio (3rd–4th century AD) shows consistent results. At Pantanello, cattle led with an absolute majority of domestic animal remains; caprines slipped to second place and surprisingly, horse rose to third place, with pig coming in fourth. Sporadic bones of ass and mule, as well as hen, were also present. In the Pantanello Kiln Deposit, the cattle, caprine, horse, and pig sequence was the same as in the previous site, with an absolute majority of cattle bones; ass and hen were also present. Thus, animal husbandry looks practically the same throughout the 6th century BC–4th century AD period of occupation.

Turning to the Roman villa site at San Biagio (late 3rd–4th century AD), a significant 58.51% majority of caprine remains could be observed among domesticates. Pig ranked in second place, while cattle and particularly horse remains were very rare. At the same time, the domestic hen, occurring here in a relatively large number for the first time in the region, nearly reached the ratio of domestic cattle in this Roman period assemblage.

It is unfortunate that the exact date of the provenience where Incoronata's only chicken bone was found is unknown. This find is of particular interest, since ten other domestic hen bones were found at 6th

Figure 1.6 Size comparison of ancient cattle: *behind*, an aurochs bull, *Bos primigenius* and *front*, a domestic ox, *Bos taurus*.

century BC Paestum (West and Ben-Xiong 1988, Table 2). A tentative explanation for this development will follow later, after the discussion of individual animal species, since some questions can only be answered through the synthetic understanding of the domestic fauna and the development of the different species.

Wild Animals

Among the wild species remains that were uncovered during the excavations in the Metaponto area, the aurochs is undoubtedly the most important and most intriguing from a zoological point of view *(Fig. 1.6)*. As prey, its killing produced by far the largest amount of meat as well as raw material of animal origin, at least three times more meat than the next largest game species, red deer. As the wild ancestor of domestic cattle, it offered an opportunity to the inhabitants of increasing the size of their domestic cattle herds. The possibility of local aurochs domestication may be hypothesized not only from prehistoric finds, but also from Roman literature. Vergil (*Georgica,* III, 532) mentions a circumstance in North Italy where, during the rule of Augustus, tamed wild aurochsen pulled a ritual wagon at a feast, because all the domestic cattle of the region had died of a plague (Lengerken 1955, 152).

A large aurochs population, consisting of beasts with very long horns, lived in the Apennine Peninsula during the Pleistocene and survived to the end of the Ice Age (Portis 1907, 110; Pohlig 1912, 38; La Baume 1947, 300ff; Lengerken 1955, 12). They became somewhat smaller and their numbers essentially diminished while their size also decreased, a well-known trend across Europe (Leithner 1927, 69; Lehman 1949, 165; Lengerken 1955, 13). Because it is a warmth-loving species, the aurochs was rarer during the Ice Age in parts of Europe north of the Alps, where the dominant large bovine was the bison, which was better suited to withstand the cold climate.

With the warming that followed the start of the Post glacial period, there was a tremendous increase in the aurochs population of continental Europe. At that time, and up to the 7th century AD, aurochsen outnumbered bison (Szalay 1915, 49; Bökönyi 1962b, 180f). This same process could not be observed in Italy, where, by contrast, the number of aurochsen seemingly decreased. The wild cattle were first probably reduced because of the pressure of the Mesolithic hunting groups, and then later forced into uninhabited areas when early agriculturalists quickly extended their cultivated land. Farmers certainly increased their domestic cattle stock with local aurochs populations,[2] but decimated wild herds.

The occurrence but also the rarity of aurochs remains could be seen in the bone samples from the Metaponto area as well. It may thus be considered natural that the animal bone assemblage from Neolithic Pantanello contained three aurochs bones. It was conspicuous, however, that neither Late Bronze Age Termitito nor Incoronata (8th–6th century BC) yielded aurochs remains, and the next bone possibly originating from an aurochs was only found at the Metaponto Sanctuary. The Greek and Roman site at Pantanello produced a fair number of aurochs bones, in spite of the dense human population during Greek times. A bone from the 2nd century AD and 11 aurochs bones from the Pantanello Kiln Deposit (2nd century BC to 1st century AD) mark the last known aurochsen in the Metaponto area. It is possible that even these bones did not originate from local wild cattle, but are results of hunting excursions into the nearby hills. Later, during Roman times, even these excursions stopped or the aurochsen became entirely extinct even in the inland areas.

Unfortunately, all aurochs bones from sites in the Metaponto area were in a rather poor state of preservation. Skull or horn core fragments could not be recovered. The few measurable bones point to medium or small sized animals that fit very well into the range of variation for aurochsen of the Apennine Peninsula. The aurochs bones brought to light at other prehistoric or early historic sites in the Apennine Peninsula were not large either, and one can therefore imagine the size decrease that often happens in small, isolated populations.[3]

Ibex, the relative of the wild Bezoar goat (*Capra aegagrus* Erxleben, 1777), the true ancestor of domestic goats, was also found at Metaponto *(Fig. 1.7)*. In

[2] Current mtDNA studies indicate that this bold statement should be treated with caution. The route of Neolithic expansion for domestic cattle from Turkey through North-Central Europe and Britain shows that the bulk of bovine mtDNA variability today derives from only a few Neolithic founder chromosomes (Bollongino et al. 2006), i. e., show negligible influence by autochtonous aurochsen in Europe. However, mtDNA indicates maternal introgression and more extensive male-mediated input may have taken place. Relevant samples from South Italy remain still to be studied.

[3] This hypothetical decrease could not be supported by statistically significant differences between the horn core measurements of Mesolithic and Late Neolithic aurochsen in Hungary (Bartosiewicz 1999).

Figure 1.7 Ibex, *Capra ibex* (L., 1758).

spite of the fact that the early Holocene distribution of ibex was much wider and its frequency was essentially higher than today, its remains are rather rare in archaeological assemblages. From South Italy its bones are known only from two unpublished sites, Rendina di Melfi (Potenza) and Grotta Scaloria (Manfredonia). One may hypothesize the prehistoric occurrence of this species in the Apennines and in North Italy's sub-Alpine and Alpine regions. In fact, Riedel (1986, 41) refers to ibex in his monograph, but only in general terms, without any specific data (site, time period, number of finds).

In the Metaponto assemblage, ibex is represented by seven remains: a left os frontale fragment with the horn core and a right distal scapula fragment from Neolithic Pantanello; a left distal scapula fragment from Termitito; a right distal scapula fragment and the shaft fragment of a left radius from Incoronata; a cervical vertebra found in a mixed Greek and Roman level at Pantanello; and a humerus fragment from the Roman period Kiln Deposit.[4]

Among these remains, the os frontale fragment with the horn core is undoubtedly the most convincing evidence for the occurrence of ibex, in spite of the fact that its damaged basal part is not measurable. The horn core is so characteristic of an adult ibex that it cannot be mistaken for that of any other ruminant species. The postcranial bones can be identified by their size and diagnostic morphological features, placing them somewhere between sheep and goat, somewhat closer to the latter. None of these pieces can be measured, though one can determine through a direct comparison with recent ibex bones that the Termitito scapula fragment originated from a large individual.

By comparison, on the other side of the Adriatic in Bosnia-Herzegovina, there are sites abundant in ibex bones. In Crvena Stijena, conspicuously large numbers of ibex bones were found (Benac and Brodar 1958, 54). Remains of this species were even more numerous in the Odmut Cave in Bosnia, where ibex was the best represented animal species in the lowest Mesolithic layer (Bökönyi, unpublished identifications). These remains showed a direct relationship with the bone industry of the Franchthi Cave and are partly related to the Balkan Mesolithic, while some other objects could be associated with the Magdalénien or with the Azilien of Southwestern and Western Europe (Srejović 1974, 5). In later layers, the ratio of ibex bones certainly decreased; however, ibex was still numerous in penultimate layer V, representing the Bubanj-Hum horizon, that is, the Late Aeneolithic (Marković 1974, 8ff). The author identified a number of ibex bones in the bone sample from Spila-Perast, a cave site on the southern Adriatic coast not far from Kotor in Montenegro. The habitation levels extended from the Early Neolithic through the Early Bronze Age. In the Franchthi Cave, ibex was found only in the Palaeolithic, and it disappeared along with Equus by the Mesolithic (Payne 1973, 59).

It is unlikely that ibex would have lived in the Metaponto area because the terrain was too flat and, with the exception of prehistoric times, was far too densely populated by humans. Today this wild goat only lives in high mountain ranges above the tree line (van den Brink 1957, 150) up to the glacier belt, and even during the winter it lives between 2300 to 3200 m above sea level (Gaffrey 1961, 208). Nevertheless, the fact that ibex remains occur at several prehistoric settlements, sometimes hundreds of kilometers from high mountains, strongly suggests that the prehistoric distribution of ibex extended over the middle-range mountains, and that the animals only withdrew from

[4] This latter specimen was not included in the original manuscript. It was noted by Erika Gál during the revision of the data base. Although no measurement is available, the humerus is a bone where size shows strong sexual dimorphism in ibex (Fernández and Monchot 2007, 489). Since the specimen was probably identified by overall size, it may belong to a buck.

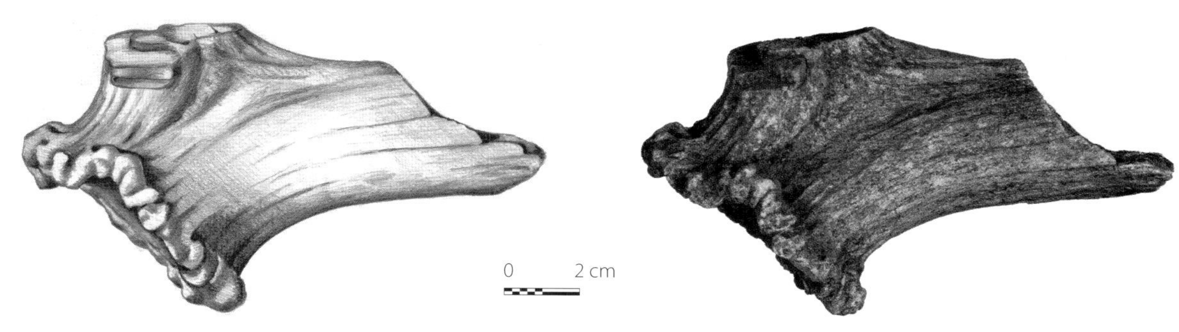

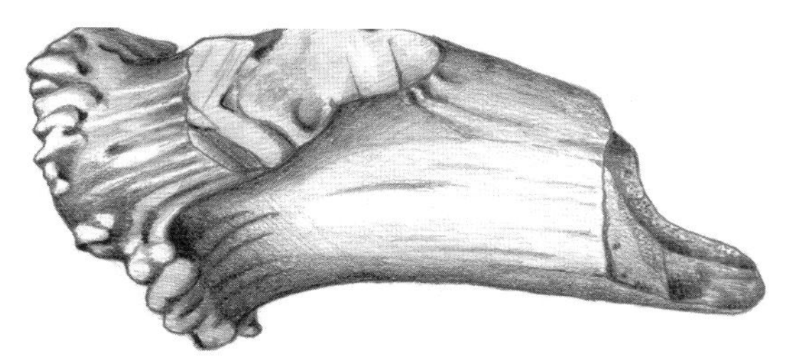

Figure 1.8 Frontal and lateral aspects of shed red deer antler burr with saw marks showing the removal of the eye tine, Pantanello Sanctuary (PZ-78-549 B).

there under the aforementioned pressure of increasing human populations and steadily expanding animal husbandry and agricultural activity. Even so, the prehistoric and early historic inhabitants had to make longer hunting excursions. But the ibex was good prey, providing quantities of fine meat since the live weight of a full-grown buck may reach as much as 110 kg (Gaffrey 1961, 208).[5]

The situation must have been similar with the chamois. Its bones (a left proximal radius fragment, a left distal metacarpus fragment, and a right metacarpus) were found only at Late Neolithic Pantanello. Chamois bones are rare at archaeological sites. They occurred in the Neolithic samples from Rendina di Melfi near Potenza and Grotta Scaloria near Manfredonia in South Italy (Bökönyi, unpublished). From Northern Italy, they were described by Riedel at sub-Alpine and Alpine sites (Riedel 1986, 41); in Serbia, they were found in the uppermost (Starčevo culture) level at Lepenski Vir (Bökönyi 1970, 1703); at Epipaleolithic Vlasac (Bökönyi 1978a, 36, 44); and at Mihajlovac–Knjepiste (Bökönyi 1992) in the Iron Gates region of the Danube. Additional finds from Bosnia are known from late Neolithic Obre II (Bökönyi 1976, 84f) in addition to Odmut Cave, where chamois remains are numerous, and Spila–Perast in Montenegro, by the southern Adriatic coast. Chamois remains have also been identified at sites in Greece: levels III and V of Sitagroi yielded a chamois horn core fragment (Bökönyi 1986a, 86f). In comparison to the Obre II metacarpus, the chamois bones found at Neolithic Pantanello were smaller, possibly owing to the less favorable and somewhat drier habitat.

Today, the preferred habitat of the chamois is high mountain forest. During the summer it moves above the tree line; in winter it descends to lower elevations (van den Brink 1957, 153; Gaffrey 1961, 211). In earlier periods, it certainly lived in medium-range hills; it later withdrew to almost inaccessible high mountains when its habitat had become increasingly taken over by human populations. Thus its story was the same as the ibex, and for its hunting, the inhabitants of Pantanello had to make similar excursions.

A considerable number of antlers were found among the red deer remains, some of which had been shed. The occurrence of shed antlers and saw marks on some of the fragments *(Fig. 1.8)* offer clear evidence that antler was a valuable raw material for tool making. Burr circumference could be measured in two shed antlers from the Pantanello complex. These measurements are 230 and 235 mm, the first from a 6th–early 5th century BC, the other from a mixed Greek and Roman layer and pointing to a larger than average size antler. At the same time, measurements taken on the fragmented extremity bones *(Fig. 1.9)* represent mainly small or even very small, or in the best case, medium-sized individuals. This can be explained by the unfavorable environment, but it is also

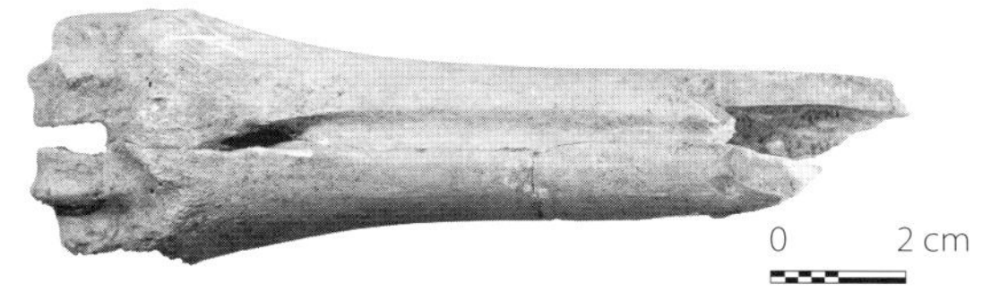

Figure 1.9 Anterior aspect of red deer metapodium, Pantanello Kiln Deposit (PZ 85).

[5] The unusually large horns may also have been a highly appreciated trophy or raw material.

Figure 1.10 Fragment of the palmate part of a fallow deer antler, Pantanello Kiln Deposit (PZ 75-168 X).

possible that the inhabitants hunted for meat and not for trophies. Therefore, there was no selection for individuals of capital sizes.

Among the rare fallow deer remains there are diagnostic antler fragments from the palmate part *(Fig. 1.10)*. Another antler fragment, found in the Pantanello Kiln Deposit, also showed the characteristics of this species *(Fig. 1.11)*. The rest of the remains consist of extremity bones and their fragments *(Fig. 1.12)*. Since the preferred habitat of this deer species is open landscape with forested spots or gallery forest along rivers, fallow deer in the area were relatively large, as is common in the Italian Neolithic. The situation was the same with the roe deer, and so it is not surprising that the measurable bones of this species originate from at least medium-sized individuals.

By contrast, the remains of wild swine indicate small or small-medium sized individuals, based on the three measurable bones: the distal width of a scapula (Termitito) at 39 mm; a humerus (San Biagio) at 42 mm; and another humerus (Pantanello Kiln Deposit) at 51 mm. These small measurements are not surprising, as the wild swine of Southern Europe generally tended to be quite small, as are even the recent wild swine in South Italy.

Figure 1.11 Fragment of the palmate part of a fallow deer antler, Pantanello Kiln Deposit (PZ 81-425).

The classification of the single cat bone, a right distal scapula fragment, is unsure. The bone was found at Bronze Age Termitito and falls within the size overlap between wild and domestic cats according to Kratochvil (1976, 30), who studied unusually large modern samples (61 domestic and 19 wild cats). He found the size range for the distal width of the scapula to be 11.7–17.4 mm in domestic cats and 15.0–18.3 mm in the wild ancestor. Thus, the 16 mm distal width of the scapula fragment from Termitito cannot be used in reliable metric identification. This bone, however, may have come from a domestic individual, since among the domestic cats of Tác-Gorsium, a Roman town in Pannonia, there were even larger domestic cats (Bökönyi 1984, 202). These latter could be the results of conscious animal breeding,[6] which could be observed in the case of other domestic animal species there. In addition, the modern reference material studied by Kratochvil consisted mainly of large Carpathian wild cats that were certainly larger than their Italian counterparts. It seems plausible therefore to suppose the wild nature of the Termitito cat, although the possibility that it was domesticated cannot be completely ruled out. The occurrence of the domestic ass, whose place and time of domestication are the same as those of the cat, may represent some indirect evidence.[7]

At any rate, one has to leave this question unanswered in hopes of finding more direct evidence in the near future. The habitat of the wild cat is any kind of landscape with enough forest or bush cover, and these elusive carnivores were not common prey items in prehistoric times.

Weasels and badgers make a small appearance in the Metaponto assemblage. Evidence for weasel comes

[6] Or even castration, as is seen in modern-day cats.

[7] Recent research has come up with alternative scenarios of cat domestication.

Figure 1.12 *Left,* fallow deer astragalus, Pantanello Kiln Deposit (PZ 81-120B-3); *right,* fallow deer metatarsus, Pantanello Sanctuary (PZ 76-452).

from late Neolithic Pantanello in the form of a single first phalanx that could be identified by the fact that this is the only weasel species on the Apennine Peninsula. Its habitat is rather variable. Weasels require bush cover and the proximity of some water. The only measurable badger bone, a left distal humerus fragment from the Pantanello Kiln Deposit, comes from a large individual, judging by its 32.5 mm distal width. The other one, a non-measurable left mandible fragment from Pantanello (2nd century AD), is of about the same size. Badger bones are rather common in prehistoric sites in Central, South, and Southeast Europe, and their flesh was certainly eaten.

Foxes in the region must have been comparatively small, as the length of a lower M1 never exceeds 14 mm. These remains fall into the lower range of size variation of prehistoric foxes in southeastern Europe (Bökönyi 1976, 151; 1986b, 131; Becker 1986, 148; etc.). Fox meat was also eaten. Like badgers, foxes burrow into the soft soil of archaeological strata and often die there. Therefore, the bones of such intrusive individuals sometimes occur in archaeozoological assemblages. It is often difficult to separate the contemporaneous from the secondary (i.e., subsequently deposited) finds of these carnivores.

The wolves in the Metaponto area were all small to medium-size animals. The incomplete skeleton found at the Pantanello Necropoleis has a whole tibia preserved *(Fig. 1.13).* Based on its 200 mm greatest length, the withers height of this animal was estimated at 60.2 cm using Koudelka's indices (Koudelka 1885), presuming that these ratios, developed for dogs, are valid for wolves as well. This is a relatively small value, since van den Brink (1957, 110) gives 70 to 80 cm as a normal withers height for a European wolf. Even if the other wolves identified may have been somewhat larger, they still seem to fall within the size range of small southern wolves. One important habitat feature for wolves is sparse inhabitation, as people have regularly killed them to protect flocks and herds.

The vole and other small rodents were almost exclusively intrusive at some sites, as is revealed by the color and state of bone preservation. The same holds true for rabbit, which did not exist on the Apennine Peninsula in the time period discussed here.

Brown hare, on the other hand, was indigenous and identified at four of the sites. All remains originated from adult individuals, displaying a small size variation. In earlier periods they did not play an essential role in human food consumption. Their importance grew in the Roman Imperial Period onwards when catching hares with greyhounds was taken over from the Celts and became a favorite pastime, not just in Italy, but also in other provinces of the Empire.

Among the wild birds remains, the most interesting were the black vulture bones found in a 4th–3rd century BC layer at the Pantanello Sanctuary *(Fig. 1.14; 4.1).* The vulture bones, along with the remains of a small bovine, were discussed by Scali (1983, 47) in connection with evidence of a sacrifice found in Rome. The special nature of the Pantanello deposit

Figure1.13 Anterior aspect of the left tibia from an incomplete wolf skeleton, Pantanello Necropoleis (Tomb 321).

Figure 1.14 Black vulture (*Ägypius monachus* L., 1758).

Figure 1.15 Re-fitted tortoise shell, Termitito (A-80-2-II-XXV).

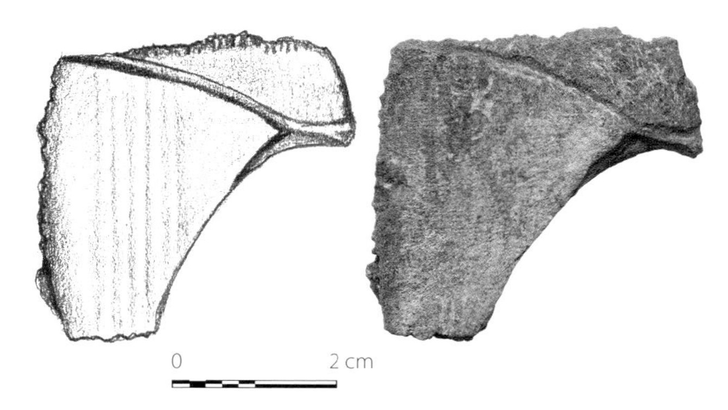

Figure 1.16 Tortoise shell fragment with cut marks, Incoronata (IC-78-350 B).

may be questioned, because black vultures and griffon vultures (*Gyps fulvus* L., 1758) were very common earlier in the southern half of Europe, and they are still present in the southwestern and southeastern regions. In the Metaponto area the vulture was a natural scavenger, whereas the occurrence of its bones in a busy urban context such as Rome might possibly be related to a special sacrifice.[8] The remaining wild bird bones belonged to ducks, but their exact identification cannot be determined because of their fragmented state.

Although the tortoise bones recovered may include remains of the pond tortoise, the majority of the shell fragments came from the Greek tortoise, a land species that is still eaten opportunistically. The difficulty posed by badger and fox bones also holds true for tortoise remains. When hibernating, tortoises dig themselves into the soft substrate of archaeological sites; because winter is a critical period, many of them die there. This is why one often finds bones or shells or shell fragments. Only cut and burn marks are definite proof that tortoise bones are synchronous with the sites under study. Such cut and burn marks are common on the tortoise bones found at sites in the Metaponto chora. On the plastron found at Termitito cut marks could be observed on two pieces that could be re-fitted *(Fig. 1.15)*.[9] These may possibly have been aimed at opening the animal's belly for the meat and extraction of the viscera inside. Conversely, it's possible that the whole shell—both plastron and carapax—could be used as a small container, which would explain keeping the fragments of the carapax together.[10] A similar cut mark could be found on a 7th–6th century BC plastron fragment from Incoronata *(Fig. 1.16)* and one from the Pantanello Sanctuary.

[8] See Gál, Chapter 4, for alternate interpretation and discussion.

[9] While several fragments of tortoise shell from Termitito indeed showed damage by human interference (cut marks and burning), subsequent studies of this particular specimen suggest a sharp, post mortem natural break.

[10] See Chapter 5 for discussion of the tortoise shell lyre found at the Pantanello Necropoleis.

Species	Latin name
Cattle	*Bos taurus* (L., 1758)
Sheep	*Ovis aries* (L., 1758)
Goat	*Capra hircus* (L., 1758)
Pig	*Sus scrofa dom. (L., 1758)*
Horse	*Equus caballus* (L., 1758)
Ass	*Equus asinus* (L., 1758)
Dog	*Canis familiaris* (L., 1758)
Cat	*Felis catus* (L., 1758)
Hen	*Gallus domesticus* (L., 1758)

Table 1.13 List of major domestic species in the Chora of Metaponto (excluding mules and hinnies).

Domestic Animals

The archaeological sites in the Metaponto area, which span a time interval of over three millennia, have yielded a bone assemblage in which all important domesticates are represented. These are shown in Table 1.13 (not including mules and/or hinnies, the bones of which were also present in the material).

Of these identified animals, cattle were the most important. The Early Neolithic was the only exception, since cattle had not yet been adapted for agricultural purposes. Even though cattle remains were outnumbered by those of caprines, the latter could not have competed with cattle in terms of meat output. After the earliest cattle domestication, it is possible people began to use milk as well. There is no direct evidence, as the first milk remains in the form of carbonized milk were found in the Hallstatt Period (Grüss 1933, 105-106);[11] the earliest representation of milking comes from a frieze at the temple of Nin-Hursag, Ur, after 2900 BC (Zeuner 1963, 219, Fig. 8.18). Still, there are several forms of indirect evidence of much earlier dairy exploitation:

1. On a representation at the temple of Nin-Hursag, cows are milked from the rear, suggesting that this technique was taken over from the milking of caprines that had been domesticated at least a thousand years earlier than cattle.[12]
2. Early domestic cows produced little milk, sufficient only for feeding their own calves. However, the mortality of newborn among the first domesticates was generally high, suggesting that people deliberately killed young calves to get milk.[13]
3. Representations from the Middle East dated to the 4th millennium BC, as well as small figurines of the middle phase of the Tripolye culture in Eastern Europe, show cattle with markedly large udders that may have been early dairy cows (Brentjes 1965, 38; Hančar 1955/56, 67).
4. There is philological evidence for the early practice of milking (J. Harmatta, personal communication). The western Indo-European languages have a common word for milking—*xmlg/xmel*— and for milk—*xmlag-ti* (actually "milking, milked")— which originate from the time prior to the separation of the western Indo-European-languages, at least from the beginning of the Neolithic (Bökönyi 1974, 109).

The use of cattle as draught animals, like their exploitation for milk, began in the Late Neolithic. Possibly first for pulling ploughs, then wheeled wagons, the draught ox became the engine of the agriculture.[14] In Roman times, cattle's dung became almost as important a product as their meat, milk, or draught power, and this also contributed to the leading role of cattle in agriculture.[15]

The prehistoric and early historic cattle population of the Metaponto area was variable both in form and size. The number of horn cores is relatively small. The

[11] Early Neolithic (5900-5500 BC) bovid milk residues have been recently identified on pottery fragments from several sites in Eastern Europe. For the time being, however, it is unclear whether they originate from sheep, goat or cattle (Craig et al. 2005). A source of confusion may be that the same vessels were used for all three types of milk as has been a widespread practice, e.g., in present-day Anatolia.

[12] This argument should be considered indirect evidence, since milking techniques did not change in a linear fashion through history. A cow being milked from the rear is also shown in a manuscript of Justinian's *Digesta* (Bologna, ca. 1340; Morrison 2007, 24).

[13] Slaughtering the newborn would not have been necessary, since the presence of the young even stimulates letdown in the dam. Early weaning would have been a sufficient alternative.

[14] The most recent summary on the subject was published in a comprehensive volume edited by P. Pétrequin et al. (2006).

[15] In the first book of Terentius Varro's *De Re Rustica*, the debate with his imaginary friends, Agrasius and Agrius, becomes heated when Varro suggests that cattle, unless used in draught work, should be excluded from further treatment. Yet cattle manure became indispensable in cultivation and Agrasius argued that it would be difficult to ignore this animal, although—according to Cassius—excrement from a variety of birds as well as human faeces were far more appreciated substances for this purpose: "*Cassius secundum columbinum scribit esse hominis*" (*RR* II, 38.15). In the case of the widespread use of cattle, it may have been the sheer quantity of manure that made these animals increasingly important for agriculture.

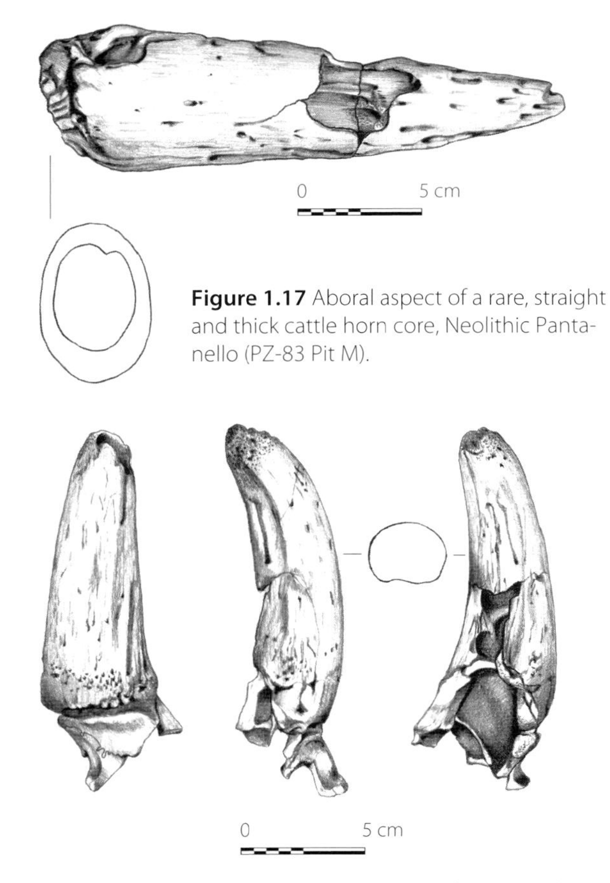

Figure 1.17 Aboral aspect of a rare, straight and thick cattle horn core, Neolithic Pantanello (PZ-83 Pit M).

Figure 1.18 Frontal, lateral and aboral aspects of a small left horn core of a cow of brachyceros type, Incoronata (IC 12).

Late Neolithic horn cores are long and heavy, with thin, furrowed walls, and with a slight or stronger torsion in most cases *(Fig. 1.17)*. Their form strongly resembles that of the wild aurochs. This makes sense as those cattle were at a low level of domestication, and almost freshly domesticated individuals may have appeared among them. A very small, fragmented horn core was found at Incoronata that would have been classified by early authors in archaeozoology as a typical brachyceros form *(Fig. 1.18)*. Large horn cores occur again in the Roman Republican period *(Fig. 1.19–20)* that also showed the typical *primigenius* type, characteristic of improved Italian cattle (Bökönyi 1984, 24). One such horn core was found along with a brain skull fragment, having the wide, flat forehead and straight intercornual ridge that are also similar to the aurochs (*Bos primigenius*).

The size of Metaponto cattle varied, as can be seen from the withers heights estimated from the greatest lengths of radii, metacarpals and metatarsals following Matolcsi's (1970) method *(Table 1.14)*. Early cattle in the Metaponto region were very small from the Neolithic through the Iron Age. The largest cow was 121 cm at the withers, while the smallest was below 114 cm, although there were some extremely large individuals in the Neolithic assemblage, as shown by two lower M3 teeth with lengths of 38 and 40 mm respectively. From North Italy, Riedel (1986, 10-11) described similar cattle from the Neolithic through the Iron Age, thus the situation was comparable.

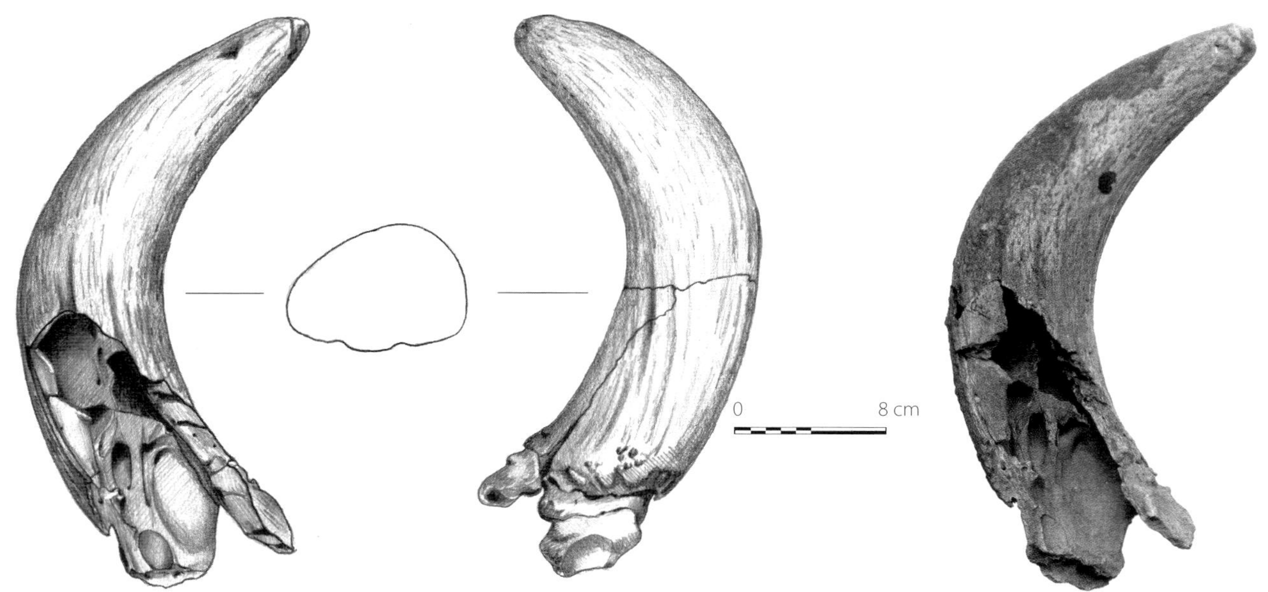

Figure 1.19 Frontal and aboral aspects of a large right cattle horn core, *primigenius* type, Pantanello Kiln Deposit (PZ-81-236).

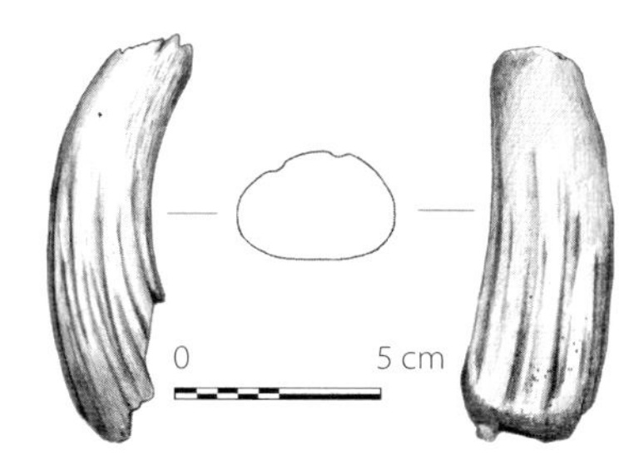

Figure 1.20 The frontal and aboral aspects of a horn core, possibly originating from an ox, Pantanello Kiln Deposit (PZ-76-585-H).

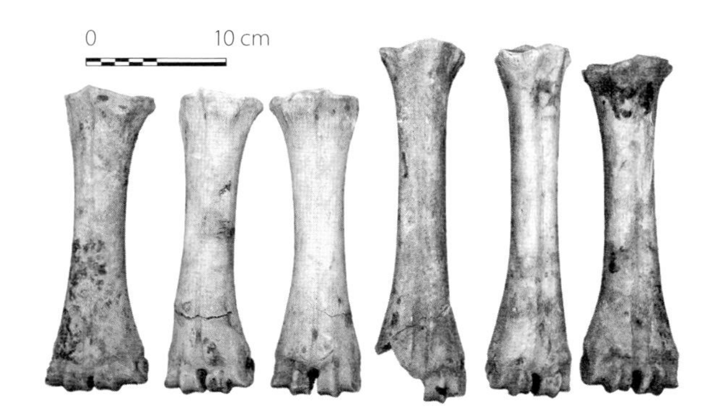

Figure 1.21 Variability in cattle metacarpals (PZ 81-462B-10, PZ 81-450B-95, and PZ 81-450B-94) and metatarsals (PZ 81-425B-1, PZ 81-450B-1, and PZ 81-425B-17), Pantanello Kiln Deposit.

With Greek colonization, larger cattle appeared in the area during the 6th century BC. Among the complete long bones, one from the Pantanello Sanctuary excelled, resulting in a withers height estimate of 127.9 cm. In addition, however, there were three very large specimens among the astragali. With a length of 77 mm each, two of these fell close to the lower limit of the size variation in wild aurochsen. A third astragalus was 72 cm long (all three came from the 5th–4th century BC period). Two distal tibia fragments reached widths of 68 and 70 mm respectively, again, very close to the line of distinction between domestic and wild cattle.

Much larger cattle were found in Roman Republican and Imperial times *(Fig. 1.21)*. The smallest individuals had a withers height of approximately 120 cm; the largest individual, evidently an ox, was over 140 cm. Even cows were over 137 and 138 cm. The average withers height of Metaponto cattle in the Roman period was 131.16 cm. In comparison, the same parameter was only 126.32 cm in the enormous sample of cattle bones from Tác-Gorsium in Pannonia (Bökönyi 1984, 28), and that was still one of the largest average withers height values calculated in the Roman provinces. Similarly high withers height values may only be found in the province of Raetia (Rüeger 1944, 236ff; Stampfli 1959-60, 436ff; Würgler 1959, 269). The smaller size of provincial cattle may have been due to the survival of individuals from the previous small, unimproved cattle population.

Bone	Greatest length (mm)	Withers height (cm)	Nobis index	Sex
metacarpus	218	134.7	28.0	F
metacarpus	219	135.3	28.8	F
metacarpus	220	136.0	-	-
metacarpus	222	137.2	30.2	F
metatarsus	220	120.3	-	-
metatarsus	228	124.7	21.1	F
metatarsus	236	129.1	24.6	M
metatarsus	245	134.0	23.1	M
metatarsus	250	136.8	22.0	F
metatarsus	253	138.4	20.6	F
metatarsus	257	140.6	21.4	M

Table 1.14 Metapodial lengths, estimated sizes, and sex determinations of cattle in the Chora of Metaponto.

M3 length, mm	36	37	38	39	40	41
Number of teeth	1		2		4	4

Table 1.15 Cattle teeth measurements, Pantanello Kiln Deposit.

Some of the Roman cattle were so large that they were very close in size to wild aurochsen (see Appendix), as can be seen in the measurement series of the length of lower M3 teeth from the Pantanello Kiln Deposit *(Table 1.15)*. The dividing line between the length of lower M3 in domestic and wild cattle is, in general, considered between 40 and 41 mm. Here the last four M3 teeth were classified as remains of domestic cattle based on osteomorphological considerations such as the fine structure of the mandibles, the gracility of teeth, etc. Similarly large cattle bones were also found in the assemblage gathered from the Villa of San Potito-Ovindoli (Early Imperial Period;

Bökönyi 1986b, 89)[16] as well as some sites in North Italy (Riedel 1974, 53; 1986, 12f).

Long-horned, large cattle were already mentioned by ancient authors, particularly Aristotle and Varro. The assemblage of cattle bones recovered from the Metaponto area shows that large, improved cattle arrived in South Italy during the Greek colonization period, although they reached larger size and probably also their characteristic constitution by Roman times. Italy, the heartland of the Roman Empire, was the center from which these animals often spread out and arrived in smaller or larger numbers in all provinces, transforming local cattle breeding.[17] And they had such a strong influence on the local cattle populations that it lasted till the Early Middle Ages (Bökönyi 1974, 138).

In addition to this large form, small cattle also lived in limited numbers in the Metaponto area in Greek and Roman times. They were survivors of the ancient autochthonous population, and they were likely the same type of cattle sacrificed at the Roman rite of the *suovetaurilia* too (Blanc and Blanc 1958-59, 42).

The sheep bones from different sites in the Metaponto area, either individually or as a group, do not show such a wide variation. This holds true even for the size variation and the small diversity of skull and horn core typology as well. As far as Late Neolithic size variation is concerned, the small sample of measurable sheep bones is indicative of medium-sized sheep of 60.3 to 64.0 cm withers heights (calculated after Teichert 1975), which are in the upper scale of the Neolithic sheep from Central and Southeast Europe (Bökönyi 1978a, 67; 1987, 140ff). Estimates calculated using unpublished sheep bones from Early Neolithic sites in Italy (such as Rendina) show that this trend is valid in the Italian Neolithic as well. During the Bronze Age at Termitito, however, a size decrease followed that was the opposite of the situation in other parts of Europe (Bökönyi 1978a, 67ff): sheep reached only 54.6 to 56.8 cm at the withers. At Incoronata, which represents both the time before the arrival of the Achaian colonists (late 8th century BC to the third quarter of the 7th century BC) and the period of their first occupation of the chora (6th century BC), sheep were somewhat larger. In addition to the estimated withers heights of 56.63 and 60.23 cm, an individual of 66.31 cm stature also appeared. It may have been an improved form, possibly imported from Greece. Finally, the only completely preserved sheep long bone from the Pantanello Kiln Deposit (2nd cen-

Figure 1.22 Reconstruction drawing of a hornless ewe on the basis of unimproved present-day sheep (Soay breed).

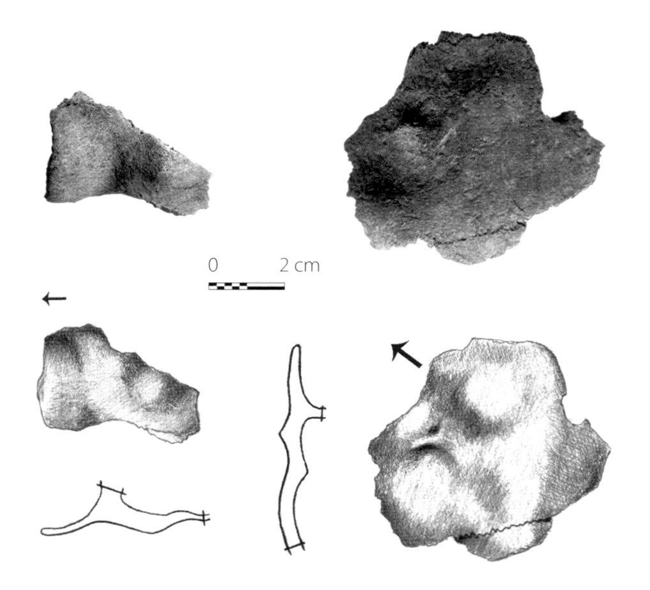

Figure 1.23 Frontal aspect and cross-section of frontal bone fragments of hornless sheep, Termitito. The arrow points in oral direction (*left*, Ter 80/2/II/XXVI.L; *right*, Ter SB1 T2).

[16] Faunal analysis at that site was completed by Erika Gál (2008).

[17] Recently, DNA studies have helped fine-tune gross theories concerning the complex reasons behind the changes in cattle size between 150 BC and AD 700 in Switzerland (Schlumbaum et al. 2003). Near Eastern mitochondrial haplotypes were found in a Roman period cattle metacarpus sample and in present-day Evolène cattle. The introduction of the Near East lineage to Helvetia must have happened during the Roman period or earlier (Schlumbaum et al. 2006).

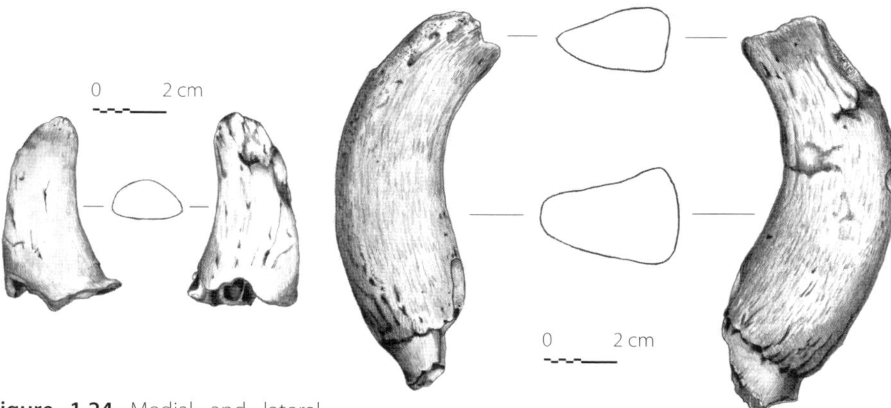

Figure 1.24 Medial and lateral aspects of a right sheep horn core of palustris type from a female, Neolithic Pantanello (PZ-83-Pit A).

Figure 1.25 Medial and lateral aspects of a large ram's right horn core of "copper sheep" type, Pantanello Kiln Deposit (PZ-76-585).

tury BC to 1st century AD) seems to have belonged to an individual that was 62.68 cm tall at the withers *(Fig. 1.3)*. However, even this value is far below the 69.54 cm average withers height calculated for the sheep from Pannonian Tác-Gorsium (Bökönyi 1984, 41). Thus the Pantanello specimen may perhaps be considered an autochthonous, unimproved animal.

This picture is consistent with the trend shown by the measurements of the extremity bone fragments. Skull fragments and horn cores, however, show a somewhat different situation. Surprisingly, hornless sheep are missing from Neolithic Pantanello. Hornlessness has widely been regarded as an early sign of domestication (Hole and Flannery 1976, Fig. 8), which first appeared in the 8th millennium and became widespread from the 7th through 6th millennia BC (Bökönyi 1977-82, Fig. 1; 1983c, Fig. 112). The absence of hornless skulls from Neolithic Pantanello may be a mere coincidence caused by the comparatively small size of the sample from this site.

The earliest remains of hornless sheep in the Metaponto area were found at Bronze Age Termitito *(Fig. 1.22)*. A small bump shows the place of the rudimentary horn cores on the cross-sections of both frontal bone fragments *(Fig. 1.23)*. At the site of Incoronata, the whole process goes one step further: on a frontal bone fragment there is a small concavity instead of the aforementioned bumps. Remains of hornless sheep occur neither at Pantanello, nor in the small assemblages from later settlements.

As for horn cores, there is not the same great variability that occurred in the large bone assemblage from Tác-Gorsium, where no less than seven types of horn cores could be described. Only two types were found here: first, a goat-horn-like, but shorter, untwisted type that can be easily identified with the so-called palustris type *(Fig. 1.24)*, which is characteristic of females, along with hornless skulls; second, a longer but thicker, twisted horn core type with triangular cross section *(Fig. 1.25)*. The latter is the horn core of male "copper sheep" types. They represent the high end of morphometric variability *(Fig. 1.26)*.

The ratio between these two types of horn cores is 4:4 at Neolithic Pantanello. Three palustris-type horn cores were found at Incoronata, and one heavy horn

Figure 1.26 A series of sheep horn cores, showing morphological variability in the material. Nos. 1–3 from Neolithic Pantanello (PZ 83 Pit A; PZ 83 Pit E; PZ 83 Pit M); nos. 4–5 from Incoronata; no. 6 from the Pantanello Kiln Deposit (PZ 76-585).

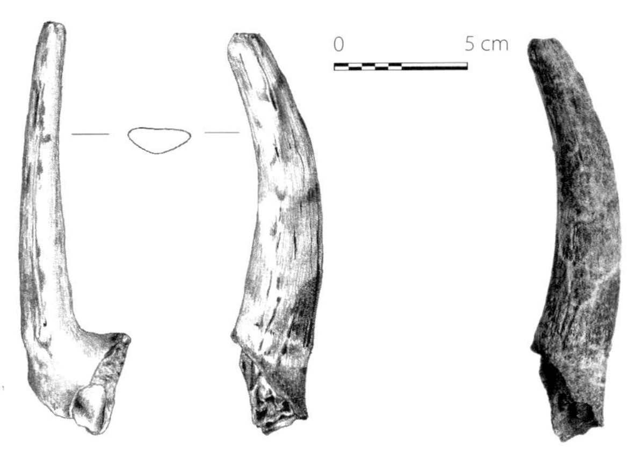

Figure1.27 Frontal and lateral aspects of a right goat horn core resembling the wild ancestor in type, Metaponto Sanctuary (PZ-75-890).

core came from Sant'Angelo Grieco. The remaining sites did not yield sheep horn core fragments of an identifiable type. Including the three hornless skull fragments, the ratio of females to males is 10 to 5, a usual proportion. In this case, however, the proportion does not have any serious implications because the horn cores came from very small samples representing different sites.

The main product of Neolithic sheep in the Metaponto region was likely meat, although dairy exploitation cannot be ruled out either. After the arrival of the Greek colonists on the southern coast of Italy, wool became the most important product from sheep. Breeds with fine fleece were imported from Greece, and they surely played an essential role in improving the local stock. At that time sheep meat and milk were not of much importance, but sheep cheese was preferred to cheese made of cow milk. Sheep were used as sacrificial animals as well (Jameson 1988, 99ff).

Remains of goats were much rarer than those of sheep at these sites. In the sample from Neolithic Pantanello, the sheep/goat ratio was 148:34 (81.31:18.69%); at Termitito, 47:2 (95.91:4.09%); at Incoronata, 47:19 (71.21:28:79%); at Pantanello 20:5 (80.00:20.00%); at the Pantanello Kiln Deposit 41:7 (85.41:14.59%).[18] Sheep/goat ratios were not counted for small sites, and neither for San Biagio, since the latter had a nearly complete goat skeleton consisting of 192 bones.

The basis for this overwhelming majority of sheep was probably a preference for its meat and—particularly in the later periods—the importance of wool. The benefits of relatively high milk production by goats (less evident in absolute quantities) could not compete with the importance of wool. As milk was also produced by cattle, goat probably became the "poor man's cow" at those times. Goat milk was also commonly used for making cheese during antiquity (Keller 1909, 303). Meat from this animal may not have been as highly appreciated, but its skin was considered as an excellent raw material.

Among the goat bones, there was only one complete metatarsal from Termitito. The 126 mm greatest length points to a large animal, measuring 67.28 cm at the withers (Schramm 1967). This animal may represent an unimproved type because its metatarsal is rather slender. As for the goat horn cores, most of them (4 specimens) are twisted outward; another two—a complete right one from the Metaponto Sanctuary *(Fig. 1.27)* and a left fragment *(Fig. 1.28)* from Incoronata—are untwisted, representing the early form that very much resembles the horn core of the wild ancestor, the Bezoar goat.

The most interesting goat find is a left frontal bone fragment from Bronze Age Termitito. This belonged to a hornless goat that is probably the earliest such individual in the prehistory of Italy. Unfortunately, the specimen is too fragmented to be shown here, even in a drawing; however, the special T-form of the coronoid suture and the typical elongated, slightly elevated bump in the place of the horn undoubtedly identify this cranial fragment as that of a goat.

In the entire bone assemblage from the chora of Metaponto, pig remains were in the worst state of preservation. One explanation is that pigs, as exclusively meat-purpose animals, were killed mainly at a juvenile or sub-adult age and their bones were less resistant and so can rarely be measured *(Fig. 1.29)*. Because they were meat animals, their bones were additionally broken up during the preparation

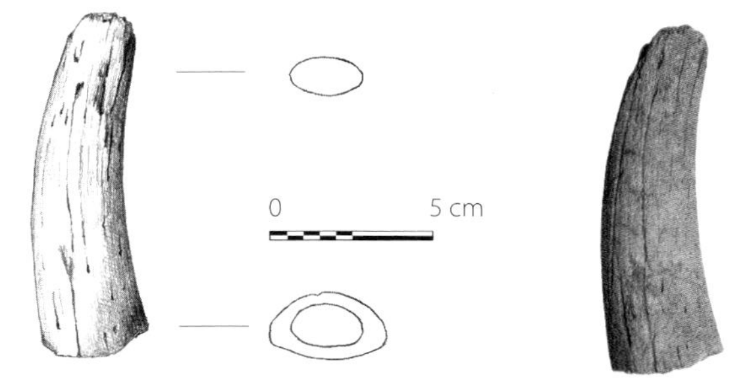

Figure 1.28 Lateral aspect of fragmented left goat horn core resembling the wild ancestor in type, Incoronata (IC-77-165B).

[18] These ratios are largely consistent with the values characteristic of archaeological assemblages across Central and Southeastern Europe (Bartosiewicz 1999, 56, Table 1).

Figure 1.29 Votive terracotta of a goddess holding a piglet, 4th century BC, from the sanctuary of Demeter at Herakleia (courtesy of the Museo Nazionale della Siritide).

of food. This limits what can be said regarding pigs at these sites. It seems that the overwhelming majority consisted of small, unimproved animals and in this respect, they did not change much, even during the period of conscious animal breeding under the Greeks and later the Romans.

The only possible sign of human influence may be at Incoronata, where there were a few larger pigs. They may also have been the result of local domestication, as suggested by the increased number of crowded premolar teeth. In the early stages of domestication, the size decrease of teeth always lags behind the shortening of the mandible. Several cases of crowded teeth were identified at this site, as can be seen in the two most typical specimens *(Fig. 1.30)*. At the same time, two mandibles with missing P1 teeth speak for a longer domesticated past in certain individuals.

The most interesting pig finds are two lower left canines from a sub-adult and a mature sow from Late Bronze Age Termitito (A 80/2/1/XXIII, h, setaggio terra, a 270), in which the tips are burnt and cracked. Another specimen, a lower canine, was found at the Pantanello Sanctuary with a partially burnt crown and cracked tooth surface *(Fig. 1.31)*. István Takács explained this type of damage as the result of singeing the relatively hairy swine of those times, and illustrated this phenomenon using modern parallels (Takács 1991). The heat in this procedure—still customary in some remote regions of Hungary and the Balkans—shrinks back the soft tissues of the mouth region, consequently exposing the incisors and canines to the direct fire, which will be more or less damaged in the way described above. The two finds are therefore interesting because they offer the first evidence for singeing pigs in prehistoric or early historic times in Italy.

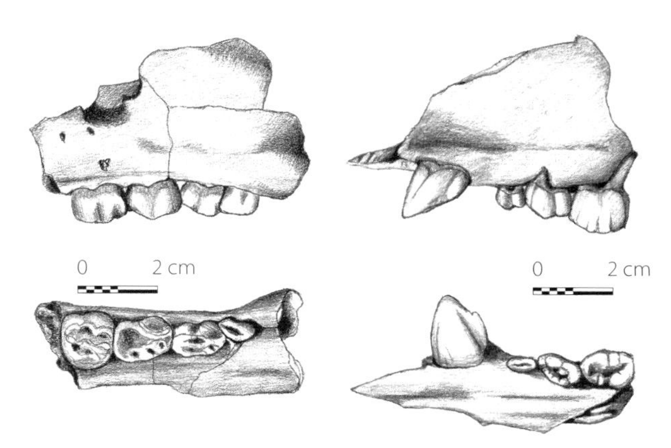

Figure 1.30 Palatal and lateral aspects of domestic pig maxillas showing crowding of upper premolar teeth, Incoronata (*left*, IC-5-Level-2); *right*, IC-78-302-B).

Among the sites in the Metaponto area, Incoronata is the only one where a relatively large number of pig bones could be sexed *(Table 1.16)*. The data show

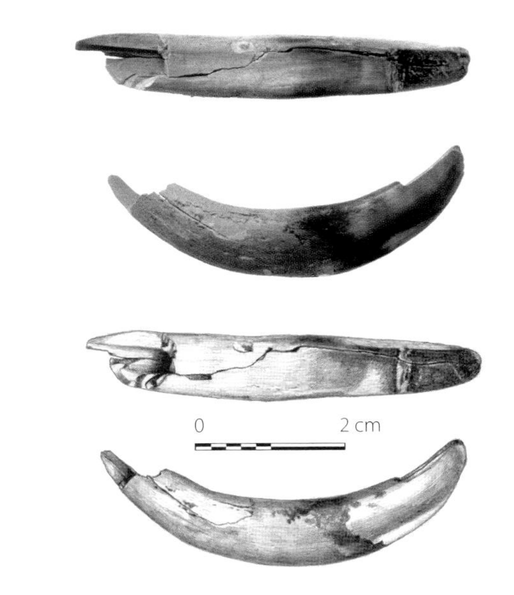

Figure 1.31 Occlusal and buccal aspects of a lower sow canine, showing probable evidence of singeing, Pantanello Sanctuary (PZ-78-444).

	Juvenile		Sub-adult		Adult		Undetermined		TOTAL
	n	%	n	%	n	%	n	%	n
Female	0	0	1	50.0	13	61.9	2	18.2	16
Male	1	100.0	1	50.0	8	38.1	9	81.8	19
TOTAL			**2**		**21**		**11**		**35**

Table 1.16 Age and sex groups of domestic pigs in Incoronata.

Neonate		Juvenile		Sub-adult		Adult/mature		Undetermined		TOTAL
n	%	n	%	n	%	n	%	n	%	n
1	1.5	15	22.7	25	37.9	23	37.9	2	3.1	66

Table 1.17 Age group ratios in the sample of pig bones at Incoronata.

surprising proportions. First, it is quite unusual that the ratio of boars is higher than that of sows; second, that both males and females were killed mainly in their adult age; and third, that the kill-off pattern for both sexes is practically the same. All these three phenomena are unknown at prehistoric sites in Central and Southeast Europe, where most of the males were killed in their juvenile and sub-adult ages (Bökönyi 1983c, 14). This theoretical kill-off pattern would have reflected that one did not have to keep as many males as females in order to sustain a "meat-purpose" species on a given level, or even to increase its stock. This is the basic impression in the bone sample from Incoronata and from other sites in the Metapontino. As was already mentioned, most pigs were killed in their juvenile and sub-adult ages, but in those age groups the sexing based on the upper and lower jaws is largely impossible, so the majority of pigs, killed at an immature age, could not be included in Table 1.16 above. According to Table 1.17, however, the bones of pigs killed at immature ages (neonate + juvenile + sub-adult) make up 62.1% of the sample, while adult/mature animals only represented a little more than one third, 37.9%. This table thus shows that the sample used for sexing was not of statistically representative size.

Bone samples from horses appeared in the latest phases at sites in the Metaponto area, even though domestic horses—the wild ancestors of which did not exist in the region in post-Pleistocene times—arrived in South Italy during the Late Bronze Age (Bökönyi and Siracusano 1987, 206). Incoronata (8 remains), the Pantanello Sanctuary (2 remains), and the Pantanello Necropoleis (a whole skeleton) were the earliest sites with horse bones in the area.

Then there were sites where horse bones were conspicuously frequent. They contributed approximately 13% of the animal remains at the stratified, predominantly Greek Pantanello Sanctuary *(Fig. 1.32)*, and 8.76% in the Roman period Pantanello Kiln Deposit. As densely settled as the Metaponto region was in Greek and early Roman times, large-scale horse-keeping somehow does not fit into the picture unless cavalry units were garrisoned there or the living standard was relatively high. The former possibility is certainly supported by the fact that several Greek poleis of Magna Graecia had strong and successful cavalry units (Azzaroli 1985, 147). In the materials recovered from agricultural villages (e.g., Sant'Angelo Grieco (6th century BC–1st century AD) or in the Roman villa at San Biagio (late 3rd–4th century AD) horse bones were far less frequent. The bone samples of both sites, however, are so small that they are statistically unrepresentative.

The most interesting horse find from all the chora excavations is an almost complete skeleton excavated at the Pantanello Necropoleis *(Fig. 3.6)*.[19] The animal was buried by itself in Tomb 316 sometime around the 4th–3rd century BC. Unfortunately, the skull and mandible were heavily fragmented. Although a metacarpus, a tibia, and a metatarsus were the only intact long bones *(Fig 1.33)*, the study of the remains yielded interesting results.

[19] See Carter (1998, 135–36, 143–44), Bökönyi (1998, 561–62), and Gál, Chapter 3).

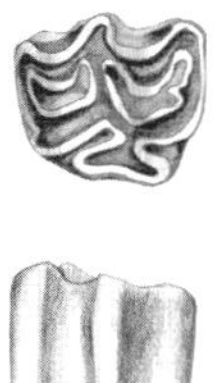

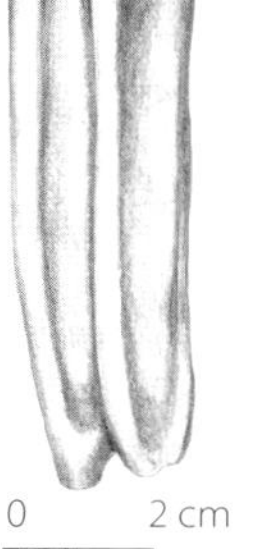

Figure 1.32 Occlusal and buccal aspects of an upper molar tooth of a horse, Pantanello Sanctuary (PZ 76 459).

First, it was the skeleton of a mature horse, because the upper and lower molar cheek teeth were extremely worn, showing spots where both the enamel and dentin were missing. The sex of this individual could not be identified because the canine teeth were missing, and its pelvis was badly fragmented. The size, however, could be estimated. Using Vitt's method (Vitt 1952, Table 1), the withers height was 146–149 cm, making this a rather large horse for the time period, and not likely a local, indigenous horse, as these were at least 20 cm lower in withers (Bökönyi 1968, 39). Originally, Greek horses were small as well, but they were probably upgraded by crossbreeding with Scythian and Persian horses. The mature horse under discussion here was surely not a common Greek horse either, since it was much larger. It was also approximately 10 cm higher at the withers than an average Scythian horse (Bökönyi 1968, 41). This tall stature points to a highly bred oriental (Persian?) horse or one crossbred with a large horse from Asia Minor.[20] A horse similar to this type is shown on a sarcophagus from Xanthus

[20] See Carter, Introduction.

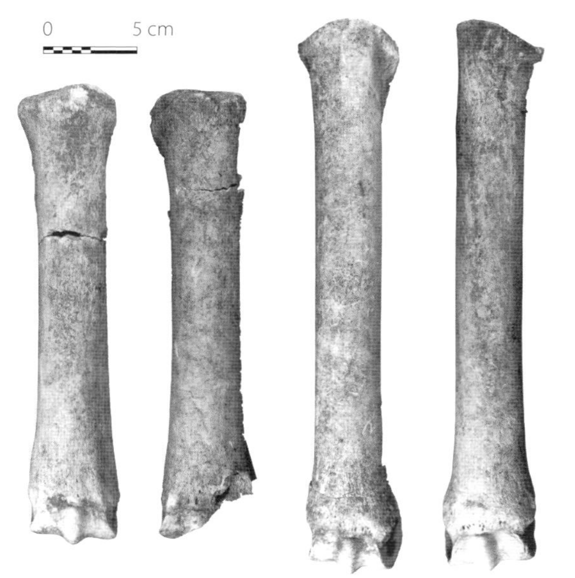

Figure 1.33 Horse metacarpals and metatarsals, Pantanello Necropoleis (Tomb 316).

Figure 1.34 3rd century BC relief from necropolis of Taras, showing charger with the same proportions as the 4th–3rd century BC horse buried in Tomb 316 at the Pantanello Necropoleis (courtesy Museo Nazionale di Taranto).

in Lycia, dated to approximately 480 BC (Anderson 1961, Pl. 13.b) as well as a 3rd century BC relief from the nearby necropolis of Taras, a colony to the east of Metaponto *(Fig. 1.34).*

The proportions of metapodia may be used to characterize the constitution (*Wuchsform*) of horses. Metacarpal measurements of the horse found at the Pantanello Necropoleis yielded a slenderness index of 24.6, which puts this individual in the middle of the range of variation for Iron Age eastern horses (Bökönyi 1968, Fig. 7), and at about the same place on the borderline between the "slightly slender-legged" and "medium slender-legged" horses in Brauner's classification (Brauner 1916). Another metacarpus from the Pantanello Sanctuary (5th–4th century BC), 232 mm at its greatest length, indicates a horse of 137.5 cm at the withers. The 14.4 slenderness index calculated for this bone points to a "slightly slender-legged horse." A third metacarpus from the same site with the same greatest length (thus the same withers height) is more robust, and its slenderness index of 15.3 places it within Brauner's "medium slender-legged" group. The first phalanges are also gracile *(Fig. 1.35),* and each of the three hoof bones (phalanx distalis) are narrow, a form thought to have been well adapted to hard steppe ground or rocky substrate.

A right maxilla fragment from Incoronata with a milk P3 or P4 also indicates a relatively large horse, with a tooth measuring 20 by 22 mm. This tooth has typical horse characteristics: a long protocone, a well-developed *pli caballini*, and rounded lateral enamel folds.

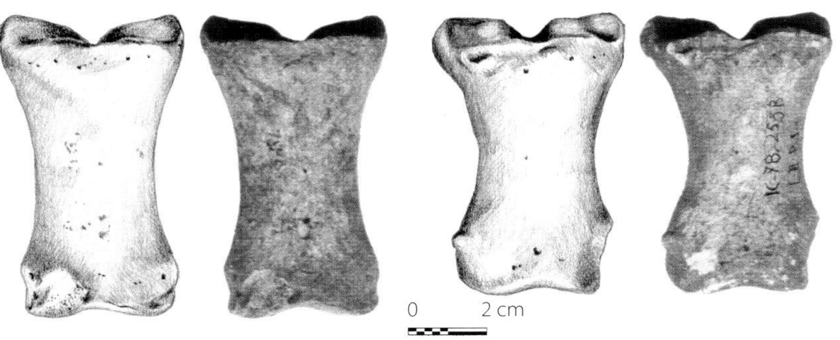

Figure 1.35 *Left*, interior aspect of horse right phalanx proximalis, Pantanello Sanctuary (PZ-82-53 BL); *right*, a horse left phalanx proximalis, Incoronata (IC-78-253 B).

Teeth of even larger horses were found at the Pantanello Sanctuary, measuring 37.5 by 25 and 37 by 26 mm respectively *(Fig. 1.36)*; a third specimen could not be measured. Non-measurable upper molars of medium-sized individuals also appeared in the 5th–4th century BC contexts in the Sanctuary and in the upper layers of the Kiln Deposit.

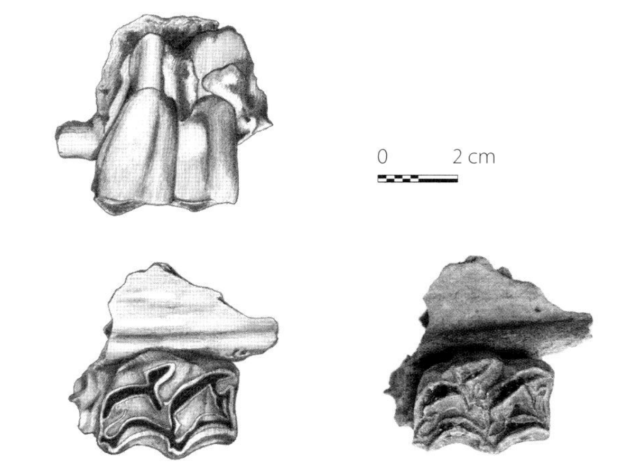

Figure 1.36 Lateral and occlusal aspects of a right maxilla with premolar tooth from a relatively large horse, Pantanello Sanctuary (PZ-78-137).

Among the horse remains from the Metaponto area, the bones from the Pantanello Kiln Deposit deserve special mention (see Appendix). They represent a comparatively short time period (2nd century BC–1st century AD) and contain numerous complete long bones *(Fig. 1.37)*, making it possible to estimate the withers height and the constitution of these horses.

The average withers height of horses was 138.1 cm in this Roman period deposit *(Table 1.18)*, which is a little higher than the average withers height of eastern (e.g., Scythian) horses, the most renowned horses in Europe outside of Greece before the Roman Empire. This average was certainly lower than that of horses

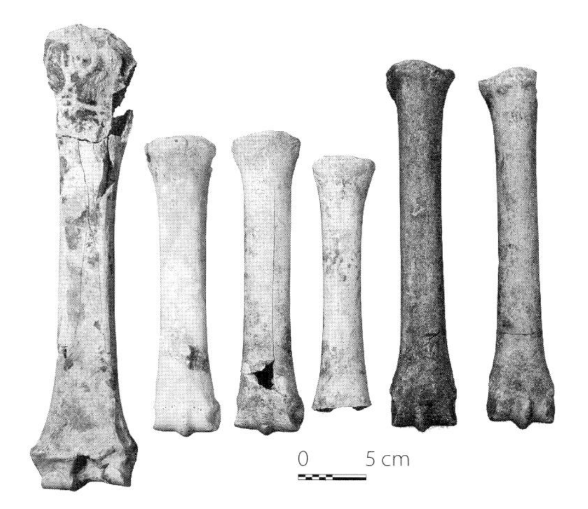

Figure 1.37 *Left,* a horse radius (PZ 81-606B-8); *right,* variability in metacarpals (PZ 81-450B-66, PZ 76-322B-6, PZ 76-309B-1) and metatarsals (PZ 81-237B-2, PZ 81-98B-2), Pantanello Kiln Deposit.

Bone	Greatest length (mm)	Withers height (cm)	Slenderness index
Radius	330	136.0	-
Radius	340	140.0	-
Metacarpus	209	130.0	14.8
Metacarpus	221	136.5	14.3
Metacarpus	226	139.0	14.6
Metacarpus	227	139.5	15.0
Metatarsus	263	137.5	11.0
Metatarsus	266	139.0	-
Metatarsus	270	140.5	10.9
Metatarsus	274	143.0	11.3

Table 1.18 Horse bone lengths, withers heights, and slenderness indexes, Pantanello Kiln Deposit.

	Juvenile		Subadult		Adult		Mature		Senile		Total
Cattle	-	-	30	16.48%	31	17.03%	115	63.18%	6	3.30%	182
Sheep/goat	1	1.11%	21	23.33%	18	20.00%	42	46.66%	8	8.89%	90
Pig	8	20.00%	9	22.50%	8	20.00%	14	35.00%	1	2.50%	40
Horse	2	3.39%	8	13.55%	9	15.25%	34	57.62%	6	10.17%	59
Dog	-	-	2	2.32%	20	23.25%	44	51.16%	20	23.25%	86

Table 1.19 Kill-off patterns of main domestic species, Pantanello Kiln Deposit.

of the later period, at least in certain provinces of the empire (Habermehl 1957, 84ff; Bökönyi 1974; 1984, 63). Among the horses at the site, only one (represented by a metacarpus of 209 mm greatest length (and 130 cm estimated withers height) may have belonged to the unimproved local breed. This supposes two forms coexisting at the settlement. While a single metacarpus cannot be seen as ample evidence for this statement, similar small and medium-sized bones from other sites support this hypothesis. Excluding this small bone, the average withers height of the improved horses at this site would be higher, 139.0 cm.

The uniformity of these horse metacarpals in terms of their slenderness is astonishing. Only one of them falls into the upper range of the "slender-legged" horses; the other three belong to the lower half of the "slightly slender-legged" group. Thus they actually represent a single continuum, and the situation is largely the same with the metatarsals. The main use of horses may have been draught power, but some were used for riding too. Horse meat was eaten in all probability, at least by certain groups of the population, as it was shown by the evidence of bone remains among the kitchen refuse of a merchant's mansion excavated in the Roman town of Augusta Raurica in Switzerland (Schmid 1970, 1317). Still, the occurrence of complete long bones from body regions of reasonably precious meat (2 radii and a tibia) suggests that horse butchery was practiced but infrequently in this area.

The data in Table 1.19 show that meat exploitation was a secondary use for the horse and unimportant in the case of dog (although, this latter species was also eaten in Augusta Raurica).[21] The only important meat-producing species was pig, with a comparatively large breeding stock (57.50% adult individuals), which means a major part of the offspring was sold either alive or as halved carcasses, salted or smoked. This type of export is hypothesized, because otherwise one would find the remains of at least five times more young individuals than adult ones. According to Keller's data (Keller 1909, 396), and presuming that half of the adult population consisted of females, a Roman sow produced on average 10 to 11 piglets every year. Cattle, sheep, and goats also had more important secondary uses than primary meat exploitation, so the absolute majority reached adult age. Nevertheless, the sheer quantity of beef was a determining factor in meat provisioning.

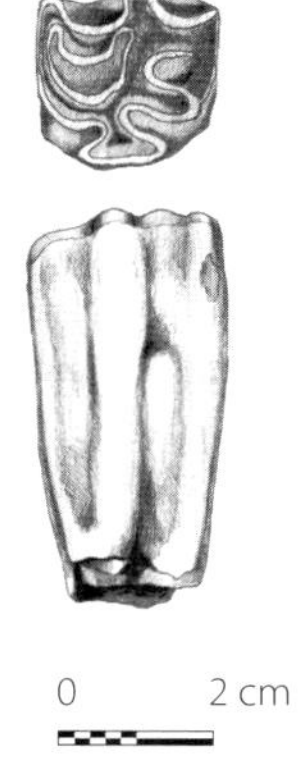

Figure 1.38 Occlusal and buccal aspects of an equid left upper molar identified as belonging to ass, Pantanello Sanctuary (PZ-75-898).

The earliest ass bone, a right pelvis fragment, was found in the Bronze Age deposit at Termitito. It did not reveal anything about the animal. In a 6th to early 5th century BC context at the Pantanello Sanctuary, a small left upper molar *(Fig. 1.38),* with lacking enamel and dentin parts, can also be put into the category of asses because of it small size, short protoconus, and the abrupt angle of transitions from the interstylar surfaces to the lateral columns (parastyle, mesostyle, and parastyle; Stehlin and Graziosi 1935, 6; Bökönyi 1972, 14f; 1986b, 307). The Pantanello Kiln Deposit also yielded six ass bone fragments, of which two mandible fragments might represent mules.

[21] There seems to be a long tradition of dog meat eating in the northeastern Alpine region and adjacent areas in Central Europe (Geppert 1990; Bartosiewicz 2006, Fig. 84).

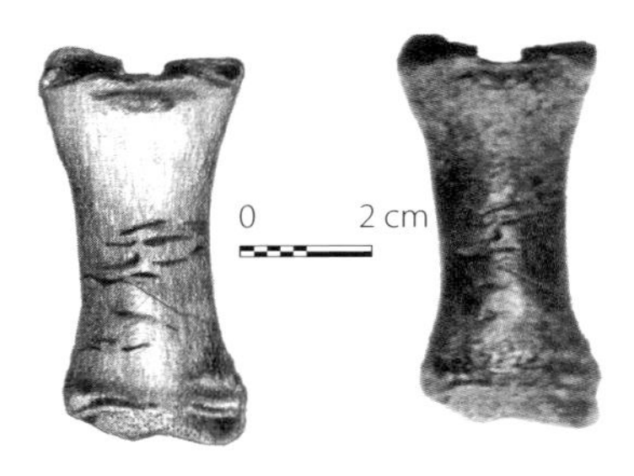

Figure 1.39 Anterior aspect of a left proximal phalanx from a hinny or mule with marks of gnawing, Pantanello Sanctuary (PZ-82-251 B).

A left anterior phalanx proximalis with gnaw marks on its dorsal surface *(Fig. 1.39)*, dated to the 1st century BC, and a left upper P3 or P4 tooth from the Archaic period are probably remains of mules or hinnies. One cannot decide their exact parentage, although there is a known representation from Oikos C at the Metaponto Sanctuary of two presumed hinnies pulling a light cart *(Fig. 1.40)*. Another smallish upper tooth can only be classified as ass/hinny/mule, as is shown when compared to a similar upper molar from horse *(Fig. 1.41)*.

In regions such as the area around Metaponto, one would have expected more asses and mules than horses. It is difficult to explain why they seem underrepresented here. One reason could be that during that early period, the area was less dry than it is today. The other is that while horses could have arrived in great numbers from the northern part of the Apennine Peninsula, asses had to be imported from Greece, North Africa, or Southwest Asia, thus expanding their stock was not an easy task.

Dogs present an interesting problem. There was no local domestication because the upper and lower tooth rows do not show any sign of crowding. Just the opposite, in three cases dogs showed such an advanced state of domestication that the premolar teeth decreased in size so that gaps occurred between them *(Fig. 1.42)*. Crowded teeth only occurred in the upper molar row, which is quite unusual and may well be the first case reported in the archaeozoological literature.

Figure 1.40 Terracotta frieze from the Urban Sanctuary at Metaponto (Oikos C) showing hinnies pulling cart in sacred procession in honor of Athena. Approximately 600 BC (courtesy of the Soprintendenza di Basilicata, photo by C. Raho).

The entire dog population seemingly belonged to a single form, however variable as a "breed." It consists mainly of large dogs, with a few medium-sized individuals and one small dog. The best preserved dog remains, the neurocranium of a young dog *(Fig. 1.1)*, points to a medium to large individual, and a typical mandible fragment shows a large but not too heavy type. The only complete long bone, a tibia from the Pantanello Kiln Deposit, has a greatest length of 190 mm. Thus the withers height estimated from it using Koudelka's method (Koudelka 1885) is 54.0 cm, which corresponds to the size of a large German shepherd. This dog was in all probability a herding or watchdog, certainly needed by the inhabitants because wolves also lived in the region.

There is little osteological variability in the dog population, that is, signs of conscious breeding are absent. The well known "luxury breeds" of the Roman Imperial Period (reminiscent of modern lap-dogs, *Dachshunds*, greyhounds, etc.; Hilzheimer 1932, 91ff; Lüttschwager 1966, 85ff; Bökönyi 1974, 320ff; 1984, 66ff; Harcourt 1974, 151; Jourdan 1976, 207ff; Hemmer and Eichmann 1977a, 268; Kokabi 1982, 81ff, etc.) are completely missing.

The earliest occurrence of domestic hen is attested by a diaphysis fragment of a tibiotarsus in the assemblage from Incoronata. Additional early domestic hen remains included a pathologically modified tarsometatarsus of a rooster from a 6th–5th century BC context at the Pantanello Sanctuary *(Fig. 1.43)*, which is discussed in the following section on pathology. A diaphysis fragment of a hen humerus was found in the 5th–4th century BC sample from Pantanello, while a later sample from the Kiln Deposit also yielded a proximal tibiotarsus fragment from this bird species. The Pantanello complex also produced a chicken femur fragment from an adult individual.

At Sant'Angelo Grieco the domestic hen was still rare, with only a single bone found. Finally, as many as 27 hen bones came to light at the 3rd–4th century AD Roman villa at San Biagio, signaling the development of a modern type of animal husbandry in which the chicken played an important part. The chicken bones at San Biagio were mainly small or medium-

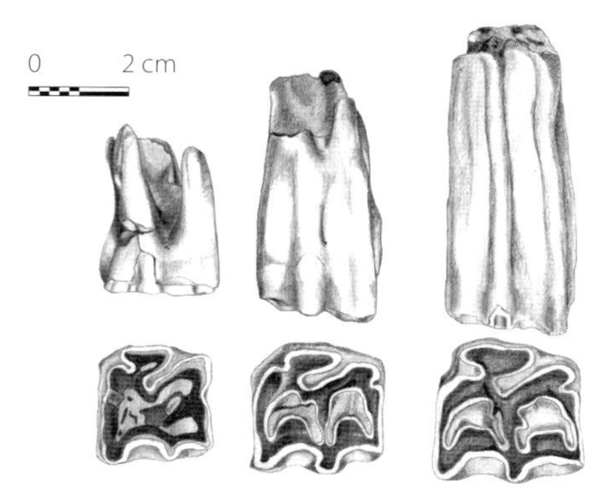

Figure 1.41 Buccal and occlusal aspects of a left upper P3 or P4 tooth (PZ-77-977) and a smallish upper tooth (PZ-77-984), classified as ass/hinny/mule, as compared to an upper molar from horse (PZ-77-1091), Pantanello.

sized, only one of them medium to large, and another indicative of a large individual. None of these birds reached the size of the larger roosters found at Tác-Gorsium in Pannonia (Bökönyi 1984, 232ff).

The sample was too small to permit the bones of the two sexes to be distinguished. Signs of caponization could not be detected either. An important use of poultry was egg production, indirectly shown by the overwhelming dominance of bones from adult individuals. Exploitation for eggs is also mentioned by the ancient sources (Zeuner 1963, 448).

Anomalies and Pathological Changes

Anomalies were not frequent at all, and pathological cases were even rarer in the animal bone samples from sites in the Metaponto area. The majority of dental anomalies may be brought into connection with the domestication that took place within a relatively short time. Among the pathological cases, fractures and arthritis were most common. These will be discussed following Wäsle's grouping (Wäsle 1976).

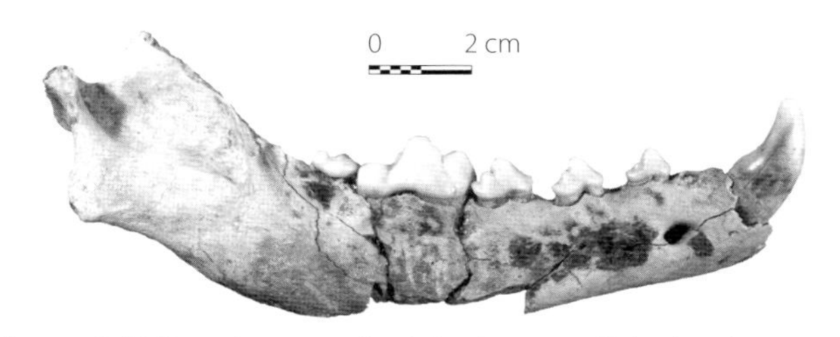

Figure 1.42 Buccal aspect of a right dog mandible showing gaps, rather than crowding, between the premolar teeth, Pantanello Kiln Deposit (PZ-81-450114).

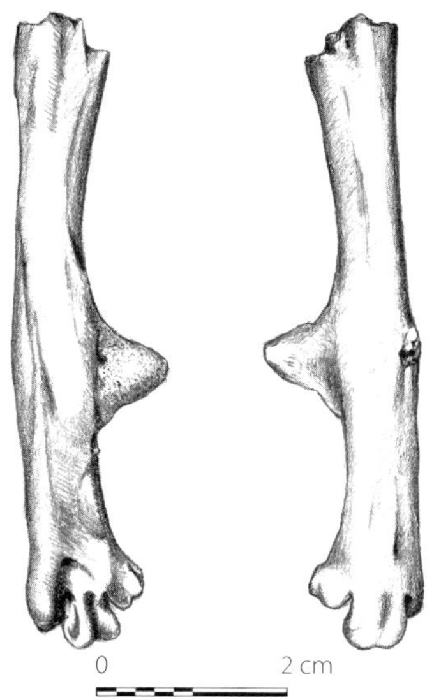

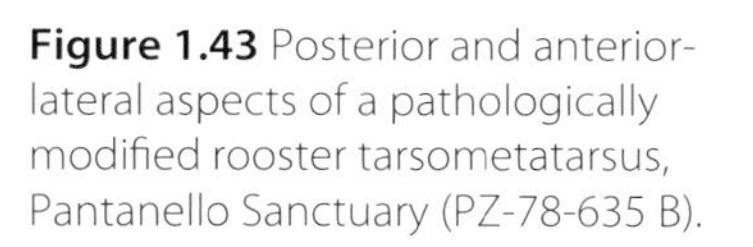

Figure 1.43 Posterior and anterior-lateral aspects of a pathologically modified rooster tarsometatarsus, Pantanello Sanctuary (PZ-78-635 B).

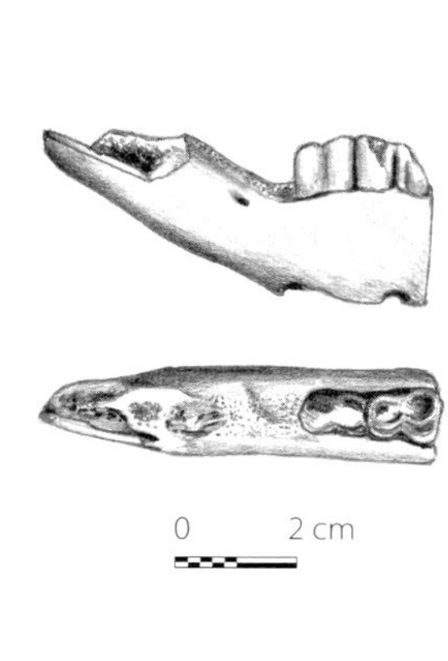

Figure 1.44 Lateral and occlusal aspects of a left mandible fragment from sheep or goat showing the evidence of in vivo loss of the P3 and P4 teeth, Neolithic Pantanello (PZ-83-Pit I 1).

Dental anomalies

Oligodonty. There were three cases of true oligodonty in this material. All of them could be observed in pig mandibles from Incoronata (8th–6th century BC). The P1 is missing, which is a symptom of recent domestication and quite natural in modern pigs.

Two other cases cannot be considered true oligodonty, although they are related to missing teeth. In a caprine mandible from late Neolithic Pantanello, the P3 and P4 fell out in vivo and their alveoli were filled in by osseous tissue *(Fig. 1.44).* In a cattle mandible from the Metaponto Sanctuary (4th century BC), the M1 was also lost in vivo. The alveolus of this tooth was almost completely filled in. Meanwhile the M2 and M3 teeth in the same mandible show enamel defects, and the M3 developed a wavy wear. These two latter symptoms result from aging.

Crowded teeth. These anomalies all tend to be related to domestication and occur in pigs and dogs alone. Three pig maxillae from Incoronata show crowded P1 and P2 teeth and the P1, P2 and P3 were all affected in a fourth specimen *(Fig. 1.30).* This is a common anomaly in freshly domesticated pigs, and it is strange that it could not be found in any of the pig mandibles, only in the upper tooth row.

The M2 and M3 teeth were crowded in a dog mandible found in the Pantanello Kiln Deposit. This

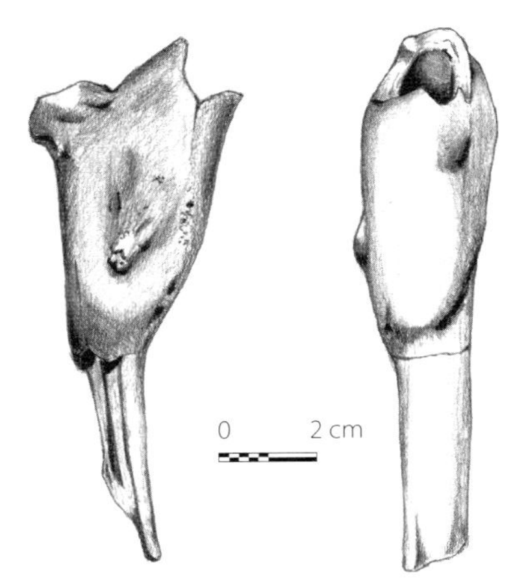

Figure 1.45 Lateral and anterior aspects of a compound fracture on a left caprine tibia. The trauma healed with grave dislocation and a heavy callus indicative of a chronic infection. Neolithic Pantanello (PZ-83 Pit 1).

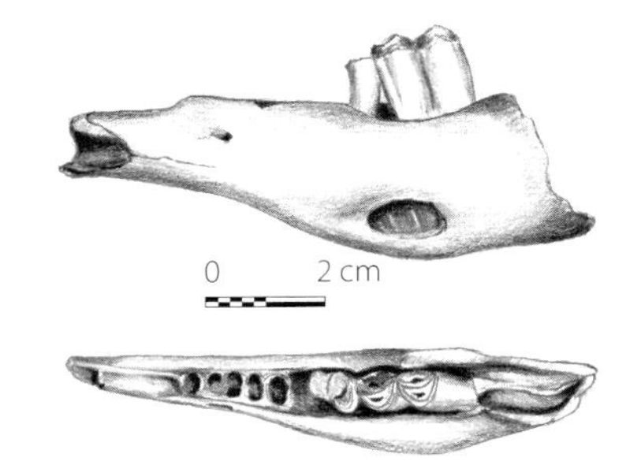

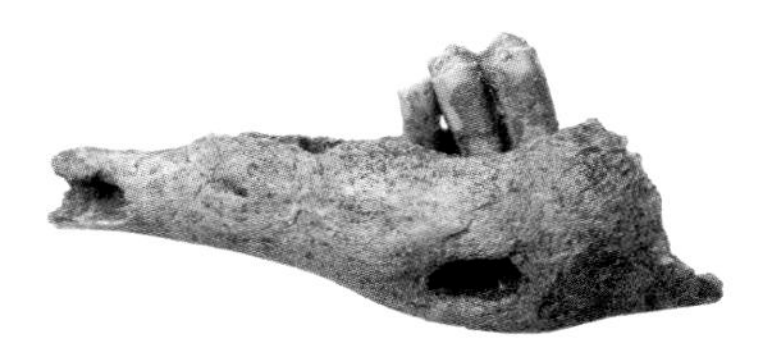

Figure 1.46 Lateral and occlusal aspects of a left mandible fragment from sheep or goat with grave symptoms of periodontitis including a fistula, Incoronata (IC-26).

is a strange type of tooth crowding because it is the premolar part of the jaw that tends to shorten with domestication and subsequently the premolars decrease in size. The reason for molar crowding may thus be something different.

Irregular tooth wear. This anomaly can be manifested in various ways and the six cases identified in the Metaponto samples represent five different types. Wavy tooth wear can be found in a caprine upper molar from the Pantanello site (6th century BC to 3rd century AD) and in another caprine left mandible fragment from Incoronata (8th–6th century BC). Sharp tooth wear was observed on the molars in a caprine mandible fragment from the Pantanello Sanctuary (late 4th–3rd centuries BC) and the aboral column of the M3 tooth in this set was overgrown. Two anomalous jaw fragments were found in the Neolithic assemblage from Pantanello: an upper tooth row with an overgrown aboral P4 and oral M1; and a lower jaw in which the M1 was worn down to its roots and the M2 and M3 show serious enamel defects. The aforementioned anomalies were all connected with the teeth material or some missing teeth. The last case was a right mandible fragment of a domestic boar from Incoronata, in which the large canine shows well developed transversal furrows. Such furrows were first discovered on the lower canines of domestic boars in early medieval Hungary (Bökönyi 1981, 99, Taf. VI. 2-4; 1974, Fig. 86). Later they were found in other regions too. The essence of this phenomenon is that as the large canines of boars did not find enough room in the shortened mandibles of the freshly domesticated pigs, they consequently became distorted and furrowed, and finally they pushed out the lateral walls of the mandibles with their roots. This phenomenon may thus be connected with the domestication of the pig.

Effects of traumatic damage and infected wounds

Fractures. Among the fractures the most interesting case is a right mandible fragment of a cow from Neolithic Pantanello with an unhealed fracture. Mandible fractures are extremely rare in cattle because of the well protected position of the jaw;[22] nevertheless, if a mandible breaks, its healing will generally be improbable because the broken parts cannot be properly fixed. Such an animal will soon die or have to be slaughtered, as happened in this case.

Another fracture from the same Neolithic site is that of a left caprine tibia diaphysis. This shows a form of malfusion where the broken ends were pulled by the muscles alongside each other. The long healing process and some possible infection caused a large callus with uneven surface *(Fig. 1.45).*

A fractured rooster tarsometatarsus, found in a 6th–5th century BC context at the Pantanello Sanctuary *(Fig. 1.43; 4.2),* had healed with a little torsion and its distal part bent forward. Finally, a cattle rib fragment from Termitito developed a false joint because the rib could not be fixed during the healing process and the constant movement hindered the reunification of the broken ends.

Periostitis. The only case of periostitis that could be observed at the Pantanello site as a whole was found on the right tibia of a horse. Large flattened exostoses grew on the anterior-medial side of the diaphysis, right above the distal epiphysis. They are probably the remains of an old trauma, maybe a kick.

[22] While no modern statistics are available for cattle, based on several decades of records kept in the Veterinary Faculty in Budapest, mandibular fractures made up only 4.88% of all fractures in horses. This value was 5.44% in dogs (Tamás ed. 1987, 44).

Arthritis. All cases of arthritis occurred on limb bones in this assemblage. A sheep radius from a 6th–5th century BC context at the Pantanello Sanctuary shows exostoses on the anterior/lateral and lateral part of its proximal articular surface. From the same site, the proximal end of a cattle proximal phalanx I exhibits the symptoms of chronic arthritis, including the extension of the articular surface by new bone formation (*Schale* in the German pathological literature). This one is a light case that did not cause lameness.[23]

On a right cattle metacarpal from the Pantanello Kiln Deposit (2nd century BC–1st century AD), several exostoses grew on the proximal end and directly below it on the anterior side. These lesions may result from chronic arthritis. The anterior side of a cattle patella from the same site shows exostoses that were probably caused by arthritis or result from simple periostitis. Finally, the distal end of a dog femur and the proximal end of the matching tibia from the same site represent a typical example of grave chronic arthritis. Their epiphyses are not only covered with exostoses but are also deformed almost beyond recognition.

Periodontitis. A very serious example of periodontitis from Incoronata can be seen in a caprine left mandible fragment. Here the molar region of the horizontal corpus swell and the spongy bone tissue doubled in thickness. When the oral column of the M1 fell out, the deep vertical abscess developed a 12 mm long and 8 mm wide fistula opening onto the lateral side *(Fig. 1.46).* This is the result of chronic inflammation caused by an infection. If the animal was a goat (this cannot be determined due to the specimen's poor preservation), it may have been a case of actinomycosis caused by microscopic fungi. These live in the oral cavity, but may infect the gum through even small wounds easily caused by hard, spiky and thorny fodder, as has been found in early domestic goats of the Kermanshah Valley, West Iran (Bökönyi 1977, 38, Fig. 35 a–b).

Chronic Deformations

Abnormalities on limb bones. The Metaponto Sanctuary (4th century BC) yielded two examples of chronic deformations, both on the distal epiphyses of cattle metatarsal bones. The distal semi-cylindrical epiphyseal condyles became considerably wider and flatter. According to von den Driesch (1975, 145) and Kokabi (1982, 128) they were the results of long lasting one-sided overloading or unnatural keeping that limits the normal motion of these animals.[24]

In sum, the number of anomalies and illnesses found on the bones from excavations in the Metaponto area are small in comparison to the size of the overall sample. This speaks for rational animal keeping with good living conditions, as well as proper feeding and treatment. It is particularly interesting that neither pigs nor dogs (two commonly affected domesticates) appeared on the list of traumatic cases.

Butchering Marks

Several of the bones recovered from the Metaponto area show butchering marks, mainly cuts. Despite this, one cannot draw general conclusions from them concerning the killing, dismemberment, and butchering of the different species, because the samples are too small and thus the bones with cut marks too few.

There is only one difference between Neolithic Pantanello and all later sites. During the Neolithic period, people used large stones before the introduction of metal axes. In other words, large bones were frequently crushed, rather than cut. Finer cuts by a small blade could only be seen in one case, which was on the neck of a brown hare scapula. From the Bronze Age onwards, heavy bronze axes were used, as demonstrated by a cattle rib, two caprine ribs, a pig ramus mandibulae fragment, and an olecranon fragment. Cut marks may be observed on a great number of tortoise carapax or plastron fragments, although in the best example *(Fig. 1.16),* all of the cuts were made using blades. In the material from Termitito, saw marks occur on a red deer antler fragment, and another antler was sawn off the skull right below the burr (see Chapter 5, Catalog No. 14).

The incidence of axe cuts increased following Greek colonization, as shown by a cattle proximal phalanx *(Fig. 1.47)* and a sheep distal scapula fragment *(Fig. 1.48)* from Incoronata. Also from this assemblage comes a dog radius, upon which two cuts were made on the medial side below the proximal epiphysis, using a knife or some other fine blade *(Fig. 1.49),* while deep cut marks were inflicted by an axe.

[23] The degree of this type of arthritic lipping is easily quantifiable and may be correlated with draught exploitation (Bartosiewicz et al. 1997).

[24] Studies on the metapodia of modern draught oxen have since shown that this deformation occurs symmetrically in both feet and may be considered one of the few reliable indicators of draught exploitation (Bartosiewicz et al. 1993).

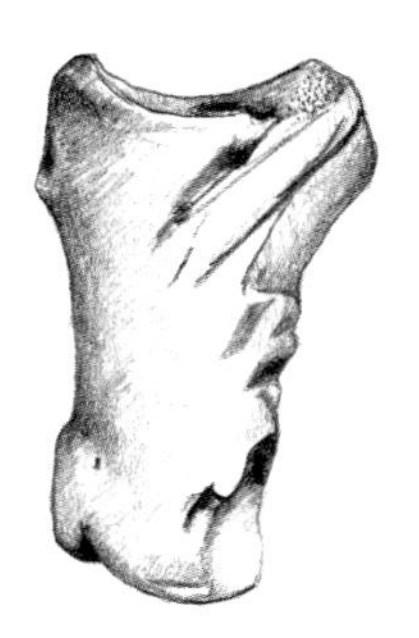

Figure 1.47 Anterio-lateral, medial and posterior aspects of a cattle proximal phalanx with heavy hack marks, Incoronata (IC-78-497 B).

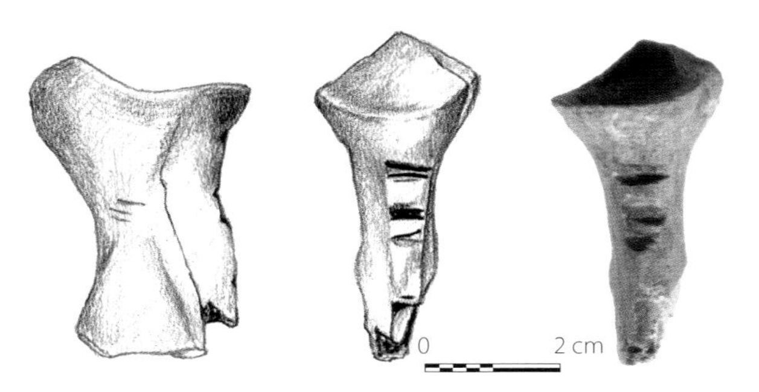

Figure 1.48 Lateral and ventral aspects, sheep distal scapula end, with fine cutting and hacking marks, Incoronata (IC-77-19 P).

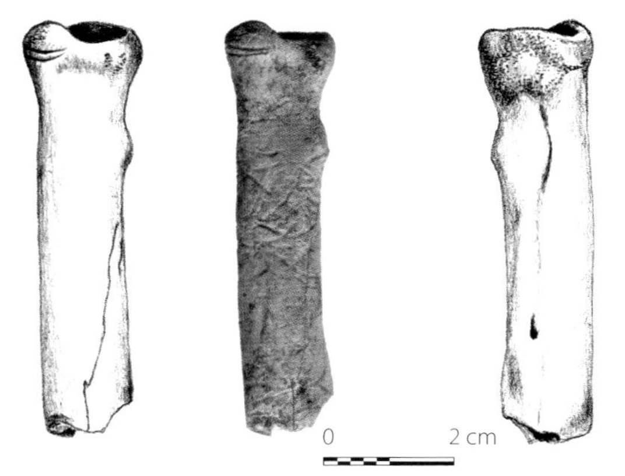

Figure 1.49 The anterior and posterior aspects of the proximal end of a left dog radius, showing fine cut marks on the medial side below the elbow joint, Incoronata (IC-77-164 P).

Bones found in the Pantanello Kiln Deposit exhibited as many as 13 such cut marks, and they occurred in great numbers elsewhere at the complex. All these chop marks were left by unsuccessful blows.

Four cattle femora, with their epiphyseal ends sawed off, were also found in the Pantanello Kiln Deposit. The aim of this procedure was to release marrow, certainly a better and more elaborate method of extraction than smashing the entire bone.[25] Moreover, the animal carcasses were not halved in the process of dismemberment: out of all samples, only a single vertebra, a cattle atlas, had been split longitudinally.

Summary

The archaeological evaluation of the animal bones excavated in the Metaponto area has resulted in a number of significant discoveries, and provided useful data not only on the faunal and early economic history of the locale, but also concerning environmental developments in the area. The sites, spanning a time period of approximately three millenia—the Late Neolithic through the Late Roman period—yielded the remains of a fauna rich both in wild and domestic species. Twenty-one wild animal taxa—16 mammals, 3 birds, 1 tortoise, and 1 fish—were found, direct descendants of Early and Middle Neolithic fauna in Italy. Nine species of domestic animals (cattle, sheep, goat, pig, horse, ass, dog, cat and hen) were represented.

The occurrence of the wild species and their habitat preferences point to the existence of four environmental types around Metaponto and its vicinity: grassland with forested spots or *Parklandschaft;* large, dense forests; gallery forests along rivers or watercourses; and medium range or high mountains. The zoological evidence suggests that the area around Metaponto was more forested in antiquity than it is today. Densely wooded areas must have existed in the region, with forest steppe and/or gallery forest habitats alternating with cultivated land.

The domestic fauna remains found at Late Neolithic Pantanello do not differ from those of the Early

[25] This thought is reversed in Chapter 2, where it is remarked that the marrow cavities were not opened. It is more likely that the saw marks represent a first step in manufacturing.

Neolithic found elsewhere in South Italy, and are indicative of strong Balkanic and Near Eastern roots. The overwhelming majority of domestic animals are caprines, followed by rare appearances by cattle and pig, and lastly dog. At Bronze Age Termitito caprines still formed the bulk of animal remains, although their ratio was much smaller. Donkey appeared as a new species in animal husbandry, along with asses in the Late Bronze Age phase of Coppa Nevigata, and this is one of the earliest occurrences in Italy. Asses may have arrived either directly from Egypt or Greece, and maybe from Asia Minor.

Animal husbandry seems to have been more successful during the Late Neolithic than in the Late Bronze Age, as evidenced by the three times higher ratio of wild animals at Termitito than in the assemblage from Pantanello.

At the pre-colonial site of Incoronata, caprines lost their absolute majority, though their bones still were the most commonly found domestic animal remains. In this period two new domestic species appeared—the horse and the domestic hen.

From the Greek colonial period (7th–6th century BC) onward, cattle took the lead among domestic animals and kept it into the Roman period, only yielding again to sheep and goats in the Late Empire, as witnessed by the site at San Biagio (late 3rd–4th century AD). The number of hen remains in the latter sample remained high, and traces of conscious animal breeding can already be recognized in the Greek Archaic period, particularly as concerned cattle and horses.

Cattle in the Metaponto area were small until the Iron Age, with withers height ranging from 114 cm to 121 cm. This situation changed suddenly in the Greek colonial period: in addition to the metacarpus of a nearly 128 cm tall bovine, several large extremity bones or fragments from large individuals appeared. Larger cattle, likely from consciously improved breeds, appeared in Roman Republican and Imperial times, when withers heights varied between 120 and 140 cm, averaging 131.16 cm. These are the large cattle that reached different provinces in the Roman Empire.

A substantial size increase could also be observed among horses, which were rather small in Italy before Greek colonization and later Roman domination under the Republic. A small horse with 130 cm withers height could be found at Metaponto as well, and presumably belonged to the old, aboriginal form. Still, in the assemblage recovered from the Pantanello Kiln Deposit, the withers height of horses already ranged between 136 and 143 cm, with an average of 139 cm. This was larger than the eastern horses of the Migration Period[26] and provided a good basis for the development of large breeds during the Roman Empire.

As far as dogs were concerned, the inhabitants of the Metaponto area enjoyed lesser variability. They succeeded to develop a large watch or herd dog, reaching the size of a modern German shepherd, but they did not have the famous Roman luxury breeds.

In conclusion, the inhabitants of the Metaponto area pursued animal keeping successfully and even conscious animal breeding. Animal husbandry provided them sufficient quantities of protein, fat, and raw materials of animal origin. To supplement their needs, or simply as pastime, people in the Metaponto region hunted a wide variety of wild animals. Still, hunting was never an essential part of their economy. In earlier times their economy was certainly based on pastoralism, as is evidenced by the high number of caprine remains. After the Greek occupation, they switched over to a more sedentary form of agriculture, in which the "engine" was the draught ox. This trend lasted until the end of the Roman Empire, when a caprine dominated form of pastoralism re-emerged, probably as a response to the reduced population and settlement. The high number of poultry remains is suggestive of the sedentism practiced by at least part of the human population.

The study of early domestic animals in the Metaponto area rests only partly on the knowledge imparted by ancient sources on animal husbandry. The archaeozoological analyses conducted in this region reveal the beginnings of conscious animal breeding, which later became widespread first in Italy, then in the provinces as well. This deliberate human intervention resulted in the emergence of specialized breeds, and increased the production of traditional domestic animals in both qualitative and quantitative respects.

[26] The Migration Period of Europe, also called the "Barbarian Invasions," is dated to between AD 300 and 700.

2
Animal Husbandry in Roman Metaponto

Sándor Bökönyi
Edited by László Bartosiewicz with Erika Gál

Editor's Note: Like Chapter 1, this is an updated, but unabridged version of the manuscript written by the late Sándor Bökönyi (1926–1994). It was singled out by the author for special analysis as a deposit perfectly suited for reconstructing Roman Period animals. Minimal editorial changes were made to the text itself. The numerical tables in this chapter were re-calculated by Erika Gál on the basis of the updated stratigraphy. Bone measurements taken after von den Driesch (1976) are included in the Appendix of the general archaeozoological review by Sándor Bökönyi.

LB

In 1975–1981, a very large midden, known as the Kiln Deposit, was excavated at Pantanello. Associated with a Roman period tile factory, the deposit produced a huge quantity of ceramic objects *(Fig. 2.1–2)*, along with coins that date the entire assemblage in successive stages from the Roman Republic period (2nd century BC) to the Early Empire (1st century AD).

Scattered throughout the deposit was a considerable quantity of animal bone, much of it in a sufficiently good state of preservation for analysis. The deposit's animal remains represent a typical bone sample from a settlement; they are mainly kitchen refuse and also contain, in smaller numbers, skeletal parts of animals that died at the settlement. Complete skeletons and skulls cannot be found among them, and even major articulated skeletal parts are extremely rare in this assemblage. Most of the bones are fragmented. Among the 1,598 identified animal remains there are only 27 long bones preserved to their full length *(Table 2.1)*.

The overwhelming majority of complete long bones (20 out of 27) came from rather meatless regions of the carcass. Only one of the seven whole long bones—a fox humerus—came from a meat rich area. The six radii represent a second-class meat region,

Figure 2.1 Excavation of Pantanello Kiln Deposit, 1980 (CW).

Figure 2.2 Mid-late 1st century BC grayware bowl with impressed cattle head design, Pantanello Kiln Deposit (PZ81.734GWf).

Species	Humerus	Radius	Metacarpus	Metatarsus	Total
Cattle	-	4	4	7	15
Sheep	-	-	2	-	2
Horse	-	1	3	4	8
Fox	1	1	-	-	2
TOTAL	**1**	**6**	**9**	**11**	**27**

Table 2.1 Complete long bones recovered from the Pantanello Kiln Deposit.

while the horse and dog bones may not be actual food remains.

Some of the bones show cut-marks inflicted during skinning or butchering. They can be divided into three types: axe cuts (e.g., on the lingual side of the ramus mandibulae of a cattle mandible and on a left caprine pelvis); knife cuts (e.g., on the distal part of a cattle scapula, on the anterior surface of a caprine radius, and on a pig tibia); and saw marks near the two ends of two cattle femora (the latter were not related to marrow extraction since the marrow cavities of neither bone were opened). Other bones—a cattle humerus and caput femoris fragment, and the anterior side of a caprine radius—show tooth marks from gnawing by dogs.

Species Types and Habitats

The animal remains identified represent a rich and varied fauna, of which eight domestic and nine wild species could be identified *(Table 2.2)*. As for domestic animals, all of the common species were present, except for cat and goose. The most important was cattle, and their bones made up 45.2% of the sample from domesticates. Caprines ranked second, followed, surprisingly, by horse and dog. Pig ranked fifth, while the bones of donkey and hen were very rare. Exotic species such as camel, peacock, and guinea fowl were missing entirely, since their importation probably followed later during the Roman Imperial era, when trade and exchange with remote regions became direct and more developed.

Vernacular name	Latin name	NISP	
Cattle	*Bos taurus* L.		641
Sheep	*Ovis aries* L.	41	
Goat	*Capra hircus* L.	7	
Sheep or goat	*Ovis/Capra*	347	
Sheep and goat	Caprinae		395
Pig	*Sus scrofa dom.* L.		108
Horse	*Equus caballus* L.		140
Ass	*Asinus asinus* L.		6
Dog	*Canis familiaris* L.		121
Hen	*Gallus domesticus* L.		7
Domestic Animals			**1,418**
Aurochs	*Bos primigenius* Boj.		11
Ibex	*Capra ibex* L.		1
Red deer	*Cervus elaphus* L.		114
Fallow deer	*Dama dama* L.		9
Roe deer	*Capreolus capreolus* L.		3
Wild swine	*Sus scrofa fer.* L.		20
Badger	*Meles meles* L.		2
Fox	*Vulpes vulpes* L.		11
Brown hare	*Lepus europaeus* Pall.		1
Tortoise	*Testudo* cf. *hermanni* L.		8
Wild Animals			**180**
TOTAL			**1,598**

Table 2.2 Fauna list with number of identifiable specimens, Pantanello Kiln Deposit.

Wild animals are represented mainly by bones of small or medium size individuals in all species. The only exception is an upper canine of a large wild boar. The most interesting wild animal remain is a nearly complete antler of a fallow deer that is rather rare in its kind. The great majority of the wild fauna consisted of the six ungulate species: aurochs, ibex, red deer, fallow deer, roe deer, and wild swine, with a significant preponderance of red deer. The larger wild ungulates were hunted for meat; still, the flesh of the two smaller carnivores—fox and badger—as well as that of hare and tortoise may have been eaten as well.

Wild animal remains provide valuable information concerning a site's environment, as the species could only have lived in a particular region if its proper habitat existed. Because red deer need large forests, woodland must have existed in the vicinity of Metaponto chora. Wild swine also live in dense forests; they can survive in gallery forests or in bushes but their preferred habitat is reed beds along watercourses. Gallery forests or grassland with forested spots were favorite habitat types of aurochs, fallow deer, and roe deer. Ibex, a mountain species, was represented by a single bone in the kiln deposit. The habitat needs of the two carnivores and the hare are not as restricted; nevertheless, they prefer a somewhat dry type of environment. This may be the case with the tortoise as well, if the eight remains indeed can be identified as Hermann's tortoise *(Fig. 2.3)*. These finds suggest that the broader area of Metaponto was more forested than it is now, and that the region's large woodland areas must have alternated with forested grassland and/or gallery forests as well as cultivated land.

Figure 2.3 Hermann's tortoise.

Animal Husbandry

The bones of domestic animals made up about 90% of the kiln deposit sample, while those of wild animals only made up the remaining 10%. Animal husbandry therefore played a far more crucial part in the livelihood of the inhabitants than hunting. Because there are no published studies of animal remains from Greek colonial or Roman settlements in southern Italy, there is a lack of information regarding the domestic fauna that existed in those periods, except what is described by classical authors (Keller 1909, 1913). This makes the animal bone sample from Metaponto extremely important, both from the viewpoint of animal husbandry practices as well as the possible creation of different breeds. While the two preliminary reports provided initial pictures of the assemblage (Cabaniss 1983; Scali 1983), they did not address the larger issues surrounding these fundamental aspects of the ancient economy.

During the working period of the kiln complex and its deposit, animal husbandry at Metaponto had certainly not yet reached the level attained in the subsequent Roman Imperial period, when the domestic hen—a leading species in modern animal husbandry—first appeared in great numbers (Bökönyi, 1974, 46). Nor had conscious selection for well-determined, independent breeds been introduced for practically every domestic species (Bökönyi, 1974, 128 ff, 178 f, 262 ff, 320 ff). In the assemblage from the kiln deposit, evidence for conscious animal breeding cannot be discerned for most of the species, such as sheep, goat, pig, ass, and hen.

At the same time, however, there are definite grounds for attributing a certain kind of conscious selective breeding in dog, horse, and cattle, which was aimed toward increasing the size of these domestic animals. The success of this effort is undisputable. For dogs, the main goal was probably to obtain large and strong working animals to serve as herding and watch dogs. The result was that the entire dog population of Metaponto consisted of medium size to large individuals, and some of them may have reached the stature of wolves. It remains an open question as to whether these large dogs are really the result of conscious selection or local domestication, but the dog remains identified apparently belonged to a single, medium to large size group. Remains of parallels to famous modern luxury breeds (lap-dogs, *Dachshunds*, greyhounds, etc.) known from the Roman Imperial period are completely missing (Hilzheimer 1932, 91

Bone	Greatest length (mm)	Withers height (cm)
radius	340	140.0
metacarpus	209	130.0
metacarpus	226	139.0
metacarpus	227	139.5
metatarsus	263	137.5
metatarsus	266	139.0
metatarsus	270	140.5
metatarsus	274	143.0

Table 2.3 Horse long bone lengths and withers heights, Pantanello Kiln Deposit.

ff; Lüttschwager, 1966, 85 ff; Hornberger 1970, 113; Bökönyi, 1974, 320 ff; 1984, 66 ff; Jourdan 1976, 207 ff; Hemmer-Eichmann 1977a, 268; Kokabi, 1982, 81 ff; Gulde, 1985, 122 ff; etc.).

A notable size increase can also be seen in horses *(Fig. 2.4)*. In Italy, small horses lived during the Iron Age as everywhere in Europe west of the Vienna-Venice line (Bökönyi, 1964, 234 ff; 1968, 19 ff). Their improvement first began at the end of the Early Iron Age, when the tribe of the Veneti imported eastern horses from the Carpathian Basin (Harmatta, 1968, 156), used them for breeding, and then sold a part of the offspring in Greece. The remains of these horses or their descendants have been recently discovered in Veneti graves (Azzaroli, 1980, 281 ff; Riedel, 1984, 236). Thus, the improvement of the Metaponto horse, shown by an increase of about 13 cm in the withers height *(Table 2.3)*, may have resulted either from horses coming from north of the Apennine Peninsula or from Greece.

The average withers height is 138.5 cm in this assemblage, which is a little higher than the average withers height of eastern horses (e. g., Scythian; Vitt 1952)—the best horses in Europe before the Roman Imperial period—although this value is certainly lower than the withers height of horses in the later period. Among the horses of Metaponto there is only one, represented by the metacarpus of 209 mm greatest length and 130 cm withers height, which could belong to the unimproved local breed. This find might imply that two breeds were living in the same place, but the occurrence of a single bone is not enough to prove this. Without this small specimen, the average withers height of the improved horses of the site would be even higher (139.8 cm).

Bones from a great variety of horses, including some exceedingly large individuals, were recovered from the excavations of Tác–Gorsium, a 1st–4th cen-

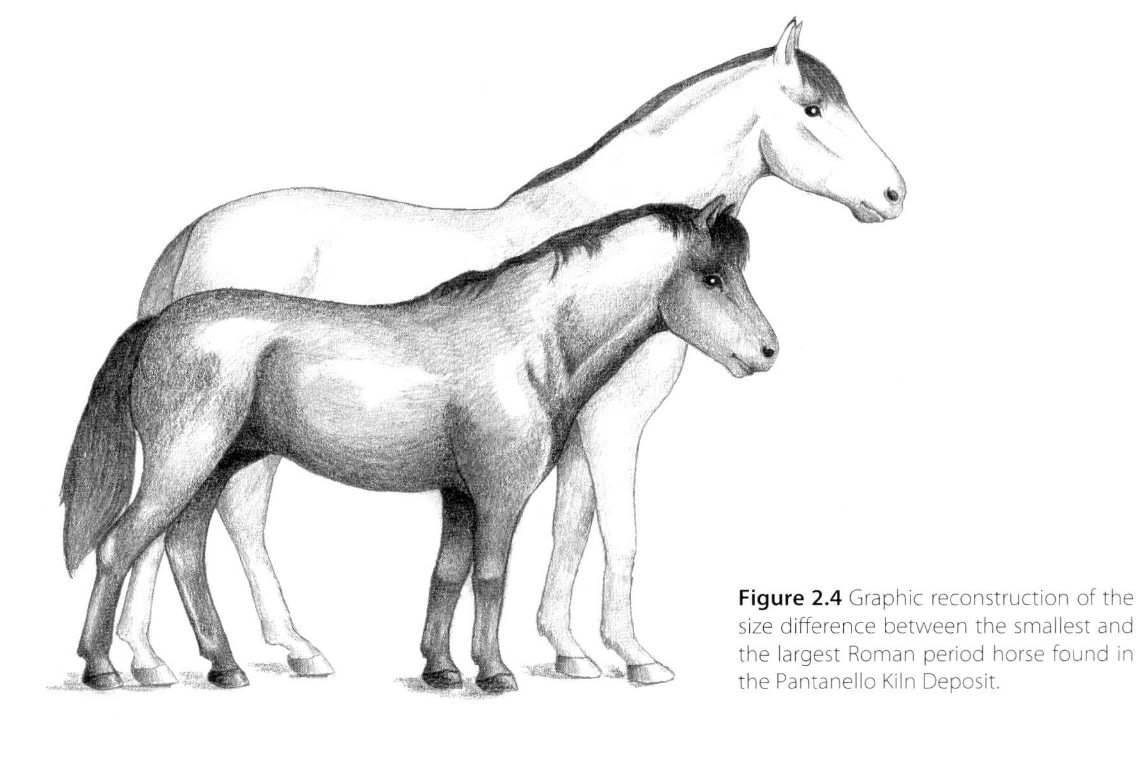

Figure 2.4 Graphic reconstruction of the size difference between the smallest and the largest Roman period horse found in the Pantanello Kiln Deposit.

tury AD Roman town in the province of Pannonia (Bökönyi 1984, 60). Four horses from Metaponto, reconstructed from the measurements of their metacarpal bones, are compared to this large group *(Fig. 2.5)*. The graph shows the estimated withers heights, in relation to Brauner's (1916) slenderness index of the bone, the smallest width of the diaphysis expressed as the percent of greatest length. The remarkably small individuals in Gorsium probably represent the horses of the local Celtic population, in general 10 cm smaller than Scythian horses (Bökönyi 1968, 36). However a few horses were unusually tall, well over 150 cm at the withers. These belong to the "slightly slender-legged" group, together with three individuals from Metaponto. Only the fourth specimen in the latter group falls within the category of "slender-legged" horses. Animals of comparable statures but with "medium slender" or even "slightly massive" legs reconstructed at Gorsium seem to indicate the genetic impact of strong Eastern horses. While the average slenderness index of western Iron Age horses averaged only 14.5%, the mean value for eastern Iron Age (Skythian) horses was 15.2% in Hungary (Bökönyi 1968, 25). This latter group, however, along with very tall horses, is missing from the Roman period kiln deposit at Pantanello.

The improvement of the local stock through conscious breeding was most successful in cattle. The cattle from this sample *(Table 2.4)* exceeded the average

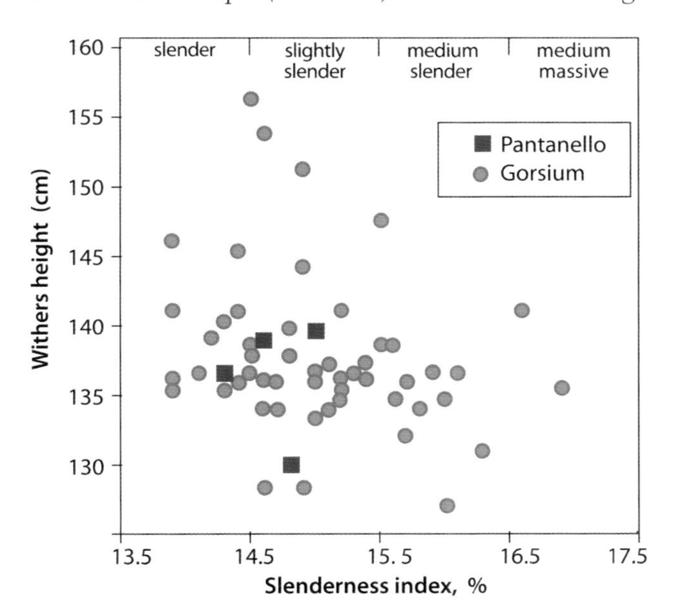

Figure 2.5 Estimated withers heights and the slenderness of metacarpal bones of Roman period horses.

Bone	Greatest length (mm)	Withers height (cm)	Nobis' index	Sex
metacarpus	218	134.7	28.0	Female
metacarpus	219	135.3	28.8	Female
metacarpus	220	136.0	-	-
metacarpus	222	137.2	30.2	Female
metatarsus	220	120.3	-	-
metatarsus	228	124.7	21.1	Female
metatarsus	236	129.1	24.6	Male
metatarsus	245	134.0	23.1	Male
metatarsus	250	136.8	22.0	Female
metatarsus	253	138.4	20.6	Female
metatarsus	257	140.6	21.4	Male

Table 2.4 Metapodial lengths, size, and sex determinations of cattle, Pantanello Kiln Deposit.

size of Roman cattle, which ranged from 124 to 126 cm at the withers (Bökönyi, 1974, 130). Their average withers height calculated using Matolcsi's method (Matolcsi, 1970, 113) was 133.5 cm. Other interesting phenomena can also be observed in Table 2.4. There is only one animal (represented by the first metatarsus in the table and having a withers height of 120.3 cm) which may belong to a primitive, unimproved breed (unfortunately, its sex cannot be determined by Nobis' method, because the proximal epiphysis of the metacarpus is damaged; if it were a cow, it could easily be a very small variant in a wide range of size variation). Another individual (with a withers height of 124.7 mm) just reaches the average size of Roman cattle; all others, however, exceed it by far. Determining the sexes using the indices developed by Nobis (1954, 179 f), cows are naturally smaller (124.7 to 138.4 cm), bulls are about the same size (129.1 and 134.0 cm), and the only ox is particularly tall (140.6 cm). These were, therefore, very large cattle, and the largest individuals nearly reached the lower limits of the range of variation in wild cattle. Nevertheless, their essentially smaller medio-lateral and anterior-posterior measurements easily distinguish them from wild cattle.

It would be interesting to know whether this highly developed cattle breeding was a local invention in south Italy in the Roman period or it had been introduced from Greece. In order to have a definite answer to this question, one would have to study large quantities of cattle remains from the Greek layers of the site and from Greece itself.

Two nearly complete horn-cores (from an adult and juvenile individual) and three horn-core fragments reveal that these cattle were long-horned and their horn form was similar to that of the aurochs. Such cattle can be found in Roman provincial sites in great numbers. Greek and Roman authors describe large, long-horned cattle: according to Aristotle, the cattle of Molossos were very large and had long horns from which drinking horns were made; Pliny mentioned the excellent cattle of Epirus; and Varro wrote that typical Italian cattle were large with a strong neck and dark or red hair coat (Keller, 1909, 332 ff). The other domestic species of Metaponto were small-sized and belonged to variable, unimproved types.

In terms of food consumption, pig was the only species which was kept exclusively for its meat. Cattle had a four-fold exploitation: milk, meat, hides, and draught power for agriculture. The first two uses were particularly important, and this is why most of the cattle were killed in their adult or mature age. Cattle would have produced about 80% of the meat consumed at the site, given that one bovine was equivalent to seven caprines or four to five pigs. The main use of sheep was their wool, but mutton was consumed too. Goats were eaten and probably also milked. Horse, donkey, and dog meat were probably avoided.

Epilogue

In a note to the original manuscript, Sándor Bökönyi stated: "The numbers of individuals have not yet been determined owing to the preliminary nature of this analysis." This reveals the fact that the rich Roman period assemblage from the Kiln Deposit at Pantanello was singled out as a sample used in characterizing the animal husbandry of that era. However, as is also shown by the wording of the introductory sentences, this text was not intended as a book chapter, but rather an independent report.

Still, this manuscript was considered worth including here, as it is a detailed study of the second largest assemblage in the chora of Metaponto, surpassed only by the 8th–6th century BC material from Incoronata. The 18 species represented by the almost 1,600 bones from this site offer a solid basis for Bökönyi's conclusions, which may be summed up as follows:

- The overwhelming majority (89%) of the bones originated from domesticates, especially ruminants, such as cattle, sheep, and goat. This is a sign of highly developed animal husbandry, well adapted to the local open grassland habitat of the coastal plain.
- Large sample size also guaranteed the representation of a broad range of species: with the exception of domestic cat and goose, all important Roman period domestic animals are present in this assemblage.
- Only a tenth of the remains originated from wild animals, dominated by cervids. The remains of red deer were far more common than those of fallow deer and roe deer. The generally small reliance on wild animals in the meat diet is understandable in a probably densely inhabited area such as the chora of Metaponto.
- Although the material consisted mostly of food refuse, it contained a sufficiently great number of long bones preserved in full length that could be used in reconstructing the withers height of several horses and cattle. Two forms of horse could be distinguished, although the smaller type was represented by only a single bone. Cattle long bones yielded consistently large withers height estimates, being indicative of a developed form.

The quantitative and qualitative characteristics of the material are indicative of highly developed Roman period animal husbandry in the area that relied on the exploitation of a broad range of domestic animals. Size variability in several species seems indicative of breed formation during this time period.

LB

3
Taphonomic Analysis of Bone Remains from the Chora of Metaponto

Erika Gál

Introduction

When Sándor Bökönyi initially studied the Metaponto materials, he concentrated primarily on the historical and faunistical importance of the assemblages. At that time, computer-based analyses were not yet common in Hungary. After Dr. Bökönyi's data had been entered into spreadsheets and were furthered examined, the idea of writing a complementary study on the taphonomic characteristics of the assemblages was formulated. The goal was to seek out the physical processes—natural and effected by humans—that followed the death of the animals, and in doing so, arrive at a new understanding of the remains.

This chapter is the outcome of two subsequent study periods and takes a taphonomic view of the materials in chronological order from the following sites: two prehistoric assemblages from Pantanello and Termitito (Late Neolithic and Late Bronze, respectively); the indigenous and later Greek settlement at Incoronata (8th–6th century BC); the Pantanello Sanctuary and the Pantanello Necropoleis (6th–3rd century BC); and three Roman sites, consisting of the Pantanello Kiln Deposit (2nd century BC–1st century AD), the farmhouse at Sant'Angelo Grieco (mainly 2nd century BC–1st century AD), and the villa at San Biagio (late 3rd–4th century AD).

The Neolithic Period and Bronze Age

The oldest assemblage from the chora of Metaponto was found at Pantanello, where Late Neolithic pits (approximately 3000–2500 BC) yielded some of the richest, most abundant materials studied. Like many Late Neolithic assemblages from Europe, remains of domestic animals were the most numerous in this group, especially sheep and goat (caprines), which constituted 81.3% of the total. The dominance of these two small ruminant species is not unique to Neolithic south Italy: sheep and goat seem to have been the most preferred domestic animal species kept at a number of Early Neolithic sites uncovered in this region, such as at Ripa Tetta, Rendina Lake 3 (phase I), Rendina, Scamuso, Torre Sabea and Favella, where the two species contributed 31.7%–68.3% of the bones to assemblages (Tagliacozzo 2005-2006: 437, Table 6). Also similar to the Pantanello pit material is an assemblage from the latest phases of a Middle Neolithic site further down the coast, Capo Alfiere, which had a preponderance of skeletal parts from sheep and goat at 68.8% and 71.9% respectively (Gál, 2008b, Table 1). The opportunistic character of hunting at both sites was evidenced by the small number of specimens related to a relatively great number of wild species. This suggests that the exploitation of animals for meat and secondary products was mainly based on a type of animal husbandry of Near Eastern origin.

As mentioned by Bökönyi (Chapter 1), the number of complete bones was relatively small in these prehistoric assemblages. This phenomenon may be explained by both the butchery customs characteristic of this earlier time, as well as by the higher degree of fragmentation due to the much greater time span. Only eight bones (0.6%) were found unbroken in the Late Neolithic assemblage from Pantanello, consisting mostly of short and compact bones such as astragali and calcanei. Among the long bones, two cattle metatarsi, a sheep radius, and a chamois metacarpus were found in a complete state or displaying only small damage *(Fig. 3.1).*

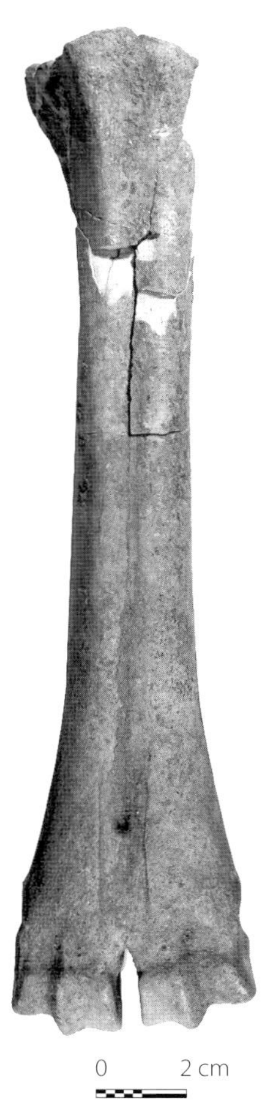

Figure 3.1 Complete cattle metatarsus with minor damage, Late Neolithic Pantanello (PZ 83 Pit H).

The anatomical distribution of remains shows that most bones originated from limbs of cattle, sheep, goat, and pig alike *(Table 3.1)*. It is worth mentioning that the skeletal representation of sheep and goat remains was the same in both the Early Neolithic levels of the Mesolithic-Neolithic transitional site of Grotta dell'Uzzo in Sicily (Tagliacozzo 1994a: 27, Table 10), and at the Middle Neolithic site of Capo Alfiere in South Italy (Gál 2008b, Fig. 6). These proportions may be related in some degree to the high contribution of these two important domesticates, i.e., their bones were deposited and found in sufficiently great numbers to better represent the skeleton; conversely, this ratio may also represent a number of taphonomic features, such as kill practices. The rather large number of meatless metapodia and phalanges indicates that the killing of caprines took place at the settlement; this applies at a much smaller scale to hare and red deer as well, which are the best represented wild animals. The presence of these bones indicates that the complete body of the animal—hunted and otherwise—was taken to the site before butchering.

The histological composition of skeletal parts, that is to say, the specific structural character, also has a significant influence on fragmentation. Teeth, the diaphyses of long bones, and the relatively short and compact bones better survive in general depositional conditions than the more fragile flat bones such as the skull, sternum, ribs, etc. Intentional breakage should also be considered: the long bones of limbs were usually split both before and after the preparation of meat. The partitioning and sharing of prey within a smaller or larger community, the secondary butchering of skeletal parts (e.g., to fit in a pot), marrow-extraction, and bone manufacturing all may have contributed to the deliberate fragmentation of bones. Half of the bone tools found in the Late Neolithic assemblage from Pantanello represent sheep and goat metapodial points (Chapter 5).

In addition to various human interventions such as butchery, cooking, and consumption, animal remains were further exposed to a number of pre- and post-depositional processes. Access by carnivorous and omnivorous animals to the kitchen midden may have also affected the abundance of assemblages and modified the distribution of certain skeletal parts. It is supposed that the small and fragile bones, as well as fragments, were usually consumed by dogs and pigs, the two main scavenging species that have lived around people for a long time. These small pieces were also the most exposed to physical and chemical destruction, such as mechanical damage by trampling, water-transport, weathering, etc. Finally, even excavation methods such as hand-collection versus wet sieving and dry screening of the excavated sediment may also have a great influence on both the faunal and anatomical distribution of skeletal parts. Large and compact bones are most likely to be recovered by hand collection.

The distribution of skeletal parts has also been analyzed in terms of the quantity and quality of meat represented by bones *(Fig. 3.2)*. Following the method described by Hans-Peter Uerpmann (Uerpmann 1973), the bones holding the finest and largest quantity of meat such as the vertebrae, the shoulder and pelvic girdles, and the proximal elements of the limbs (humerus and femur) have been grouped to form Category A. The medium value meat (Category B) is connected to skeletal parts holding less edible tissue, such as the brain skull, mandible, ribs, and the middle elements of limbs. The rest of the bones—e.g., maxilla, tail vertebrae, and distal leg bones—represent the lowest value meat (Category C). Although favoring or excluding certain body parts from human consumption may be characteristic to certain cultures, this method based on a logical categorization both from a quantitative and qualitative point of view is largely used among archaeozoologists when analyzing meat distribution based on skeletal remains.

Study of the prehistoric site at Pantanello indicated a slight predominance of remains representing Category B meat in all four of the well-represented domestic species. Since a portion of the relevant skeletal parts—e.g., skull and ribs—are more fragile, they contribute many fragments to this category. While the long bones are often split, this high proportion of Category B bones does not suggest any unusual feature. The number of almost entirely meatless parts from Category C again suggests that these domesticates must have been killed and prepared within the excavated part of the settlement. This contrasts with the predominance of Category C elements in the Middle Neolithic assemblages from Capo Alfiere, where the higher number is likely related to the finer methods of recovery used during the excavation. The more refined collecting methods of wet sieving and

Skeletal part	Cattle	Sheep and goat	Pig	Dog	Aurochs	Ibex	Chamois	Roe deer	Wild boar	Wild ass	Weasel	Fox	Hare
horn core/antler*	1	21											
neurocranium	4	49				1							
viscerocranium	1	5	2										
maxilla	2	14	1										
dentes	15	101	9						1				
mandibula	22	66	6	1								1	
Head Subtotal	45	256	18	1	0	1	0	0	1	0	0	1	0
atlas		1											
axis		4											
vert. cervicalis	5	15	2										
vert. thoracalis		17											
vert. lumbalis	1	6											2
os sacrum		2											2
pelvis	3	43	1					1				1	7
vert. caudalis		1											
sternum	3												
costa	23	144	20							1			4
Trunk Subtotal	35	233	23	0	0	0	0	1	0	1	0	1	15
scapula	7	46	9		1	1			1				1
humerus	5	62	5		1								1
femur	8	38	5	1									1
patella	1	1											
radius	5	75	4		1		1		1				4
ulna	1	18	4						1				1
tibia	4	116	2										2
fibula			2	1									1
carpalia	4	11											
metacarpalia	3	35	1				2						1
sesamoideum													
calcaneus	1	16	1										1
astragalus	4	16							1				
centrotarsale	1	6											
metatarsalia	6	52	7										5
ph. proximalis	10	27	6						1		1		1
ph. media	4	19											
ph. distalis	3	11	1						1				
Limb Subtotal	67	549	47	2	3	1	3	0	6	0	1	0	19
Total	**147**	**1038**	**88**	**3**	**3**	**2**	**3**	**1**	**7**	**1**	**1**	**2**	**34**

Table 3.1 Anatomical distribution of remains by species, Neolithic Pantanello. *Only antler attached to skull. Red deer frags. not included.

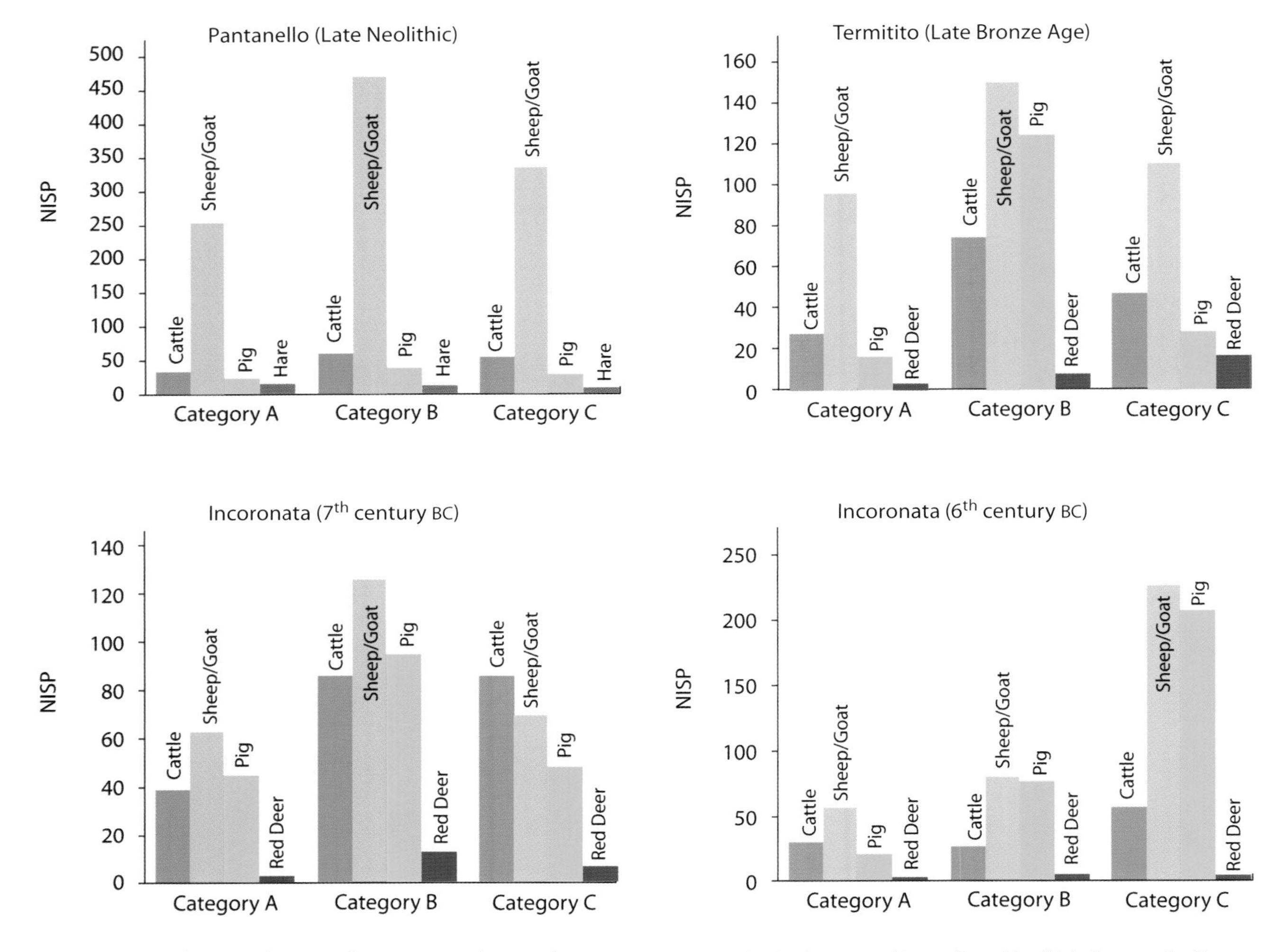

Figure 3.2 Distribution of remains by species and strata showing meat categories in the assemblages from Neolithic Pantanello, Bronze Age Termitito, and Archaic Incoronata. Category A = highest value meat; Category B = medium value meat; Category C = lowest value. NISP = number of identified specimens.

dry screening the sediment also resulted in a greater number of small but identifiable remains including teeth, metapodial diaphysis fragments, phalanges, etc., all representing body parts poor in meat (Gál 2008b, Fig. 6).

Owing to the great fragmentation of remains from Pantanello, only a small part of the material could be aged. The age distribution *(Table 3.2)* shows that among ruminants mostly adult and mature animals were slaughtered. This distribution points to the secondary exploitation of cattle (milk and draught cattle) and caprines (milk and wool), already a common trend during prehistoric times. Among the swine population, piglets were slaughtered in greater numbers than adults, much in contrast to calves and lambs. Most possibly this was due to the fact that adult pigs have no form of secondary exploitation, so relatively few were kept alive until maturity for the purposes of stockbreeding. It must be noted, however, that this species contributed less than 7% to the assemblage. The same age distribution of remains was found regarding ruminants both in the earlier and later Middle Neolithic materials of Capo Alfiere. Remains from adult pigs were identified in a greater quantity than those of piglets, but the small number from young animals may also be due to the high degree of erosion the more fragile bones experienced in that material. Nevertheless, that whole assemblage similarly suggested the secondary exploitation of domestic ruminants beyond meat provisioning (Gál 2008b, Table 3).

Table 3.2 *(Facing page)* Age distribution in domestic and most frequent wild species by sites and periods.

Site	Species	Juvenile	%	Subadult	%	Adult	%	Mature	%	Senile	%	Undeter.	%	Total	%
Pantanello (Late Neolithic)	Cattle	0	0	8	5.44	11	7.48	18	12.24	0	0	110	74.82	147	100.00
	Sheep and goat	8	0.77	76	7.32	116	11.17	124	11.94	7	0.67	707	68.11	1.038	100.00
	Pig	10	11.36	8	9.09	9	10.22	6	6.81	1	1.13	54	61.36	88	100.00
	Dog	0	0	0	0	0	0	1	33.33	0	0	2	66.66	3	100.00
	Hare	0	0	0	0	0	0	21	61.76	0	0	13	38.24	34	100.00
Termitito (Late Bronze Age)	Cattle	0	0	4	2.70	3	2.03	15	10.13	2	1.35	124	83.78	148	100.00
	Sheep and goat	2	0.61	28	8.56	34	10.40	23	7.03	0	0	240	73.39	327	100.00
	Pig	1	1.03	14	14.43	15	15.46	8	8.25	2	2.06	57	58.76	97	100.00
	Dog	0	0	0	0	1	11.11	6	66.66	0	0	2	22.22	9	100.00
	Red deer	0	0	0	0	3	8.10	4	10.81	0	0	30	81.18	37	100.00
Incoronata (8th c. BC)	Cattle	0	0	10	4.76	4	1.90	41	19.52	2	0.95	153	72.85	210	100.00
	Sheep and goat	0	0	7	2.67	20	7.63	47	17.93	2	0.76	186	70.99	262	100.00
	Pig	1	0.53	13	6.91	25	13.29	23	12.23	2	1.06	124	65.95	188	100.00
	Horse	0	0	0	0	2	50.00	2	50.00	0	0	0	0	4	100.00
	Dog	0	0	0	0	0	0	3	37.5	0	0	5	62.50	8	100.00
	Red deer	0	0	2	5.55	0	0	4	11.11	0	0	30	83.33	36	100.00
Incoronata (6th c. BC)	Cattle	0	0	1	0.88	4	3.50	22	19.30	3	2.63	84	73.68	114	100.00
	Sheep and goat	0	0	13	3.54	25	6.81	102	27.79	35	9.53	192	52.31	367	100.00
	Pig	1	0.32	4	1.31	27	8.85	119	39.01	6	1.96	148	48.52	305	100.00
	Horse	0	0	0	0	0	0	1	33.33	2	66.66	0	0	3	100.00
	Dog	0	0	1	25.00	0	0	2	50.00	0	0	1	25.00	4	100.00
	Red deer	0	0	0	0	0	0	2	11.76	0	0	15	88.23	17	100.00
Pantanello-Sanctuary (6th–3rd c. BC)	Cattle	0	0	24	4.34	33	5.96	81	14.64	30	5.42	385	69.62	553	100.00
	Sheep and goat	6	2.64	7	3.08	10	4.40	30	13.21	3	1.32	171	75.33	227	100.00
	Pig	2	3.45	2	3.45	6	10.34	9	15.51	2	3.45	37	63.80	58	100.00
	Horse	0	0	7	4.73	12	8.10	60	40.54	10	6.75	59	39.86	148	100.00
	Ass	0	0	0	0	0	0	4	50.00	4	50.00	0	0	8	100.00
	Dog	0	0	3	9.68	0	0	20	64.51	0	0	8	25.80	31	100.00
	Red deer	0	0	1	1.09	0	0	12	13.18	2	2.19	76	83.51	91	100.00
Pantanello-Kiln Deposit (2nd c. BC–1st c. AD)	Cattle	0	0	30	4.68	31	4.83	115	17.94	6	0.93	459	71.60	641	100.00
	Sheep and goat	1	0.25	21	5.31	18	4.55	42	10.63	8	2.02	305	77.21	395	100.00
	Pig	8	7.40	9	8.33	8	8.33	14	12.96	1	0.92	68	62.96	108	100.00
	Horse	2	1.42	8	5.71	9	6.42	34	24.28	6	4.28	81	57.85	140	100.00
	Ass	0	0	0	0	0	0	0	0	1	25.00	3	75.00	4	100.00
	Dog	0	0	2	1.65	20	16.52	44	36.36	20	16.52	35	28.92	121	100.00
	Red deer	0	0	2	1.75	1	0.87	25	21.92	2	1.75	84	73.68	114	100.00
Sant'Angelo Grieco (2nd c. BC–1st c. AD)	Cattle	0	0	1	1.64	3	4.92	13	21.31	0	0	44	72.13	61	100.00
	Sheep and goat	0	0	2	13.33	1	6.66	2	13.33	0	0	10	66.66	15	100.00
	Pig	0	0	1	12.5	1	12.5	1	12.5	0	0	5	62.5	8	100.00
	Horse	0	0	0	0	0	0	0	0	0	0	6	100.00	6	100.00
	Dog	0	0	0	0	0	0	0	0	0	0	2	100.00	2	100.00
	Red deer	0	0	0	0	3	10.71	6	21.42	0	0	19	67.85	28	100.00
San Biaggio (3rd c. BC–4th c. AD)	Cattle	0	0	0	0	4	12.90	1	3.22	0	0	26	83.87	31	100.00
	Sheep and goat	1	0.30	3	0.92	199	61.60	10	3.09	1	0.30	109	33.74	323	100.00
	Pig	36	22.08	15	9.20	21	12.88	15	9.20	0	0	76	46.62	163	100.00
	Horse	0	0	0	0	0	0	1	25.00	0	0	3	75.00	4	100.00
	Dog	0	0	0	0	0	0	1	50.00	0	0	1	50.00	2	100.00
	Hen	0	0	0	0	1	3.44	7	24.13	7	24.13	14	48.27	29	100.00
	Red deer	0	0	0	0	0	0	0	0	0	0	3	100.00	3	100.00

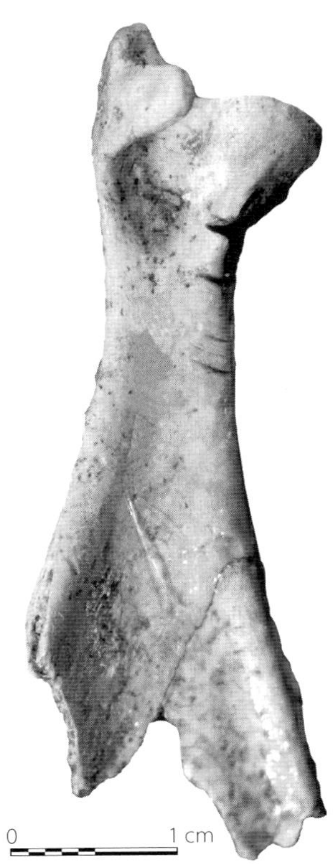

Figure 3.3 Hare scapula with parallel cut marks on the collum scapulae. Late Neolithic Pantanello (PZ 83 Pit JJ).

In spite of the high degree of fragmentation of remains at Pantanello, only two bones displayed cut marks. This may be due to a special butchery technique wherein most of the meat was first removed, and the bones were only grooved and possibly damaged afterwards. Since metal knives were not yet available in the Neolithic period, cutting up hard materials with stone blades must have been rather difficult; moreover, the great number of rather small skeletal parts from caprines did not require much fragmentation. Interestingly, a scapula from hare found in this assemblage displayed several parallel cut marks on the collum *(Fig. 3.3)*. In addition, 35 burnt remains (2.6% of the assemblage), including the long bones and carapace fragments of tortoise, show contact with fire. Although several remains at the site could have been burnt secondarily (already as food refuse) and not necessarily during food preparation, the body parts involved suggest that these bones and the attached meat were roasted. One also would expect more remains displaying massive fire marks if these bones were charred as a part of fire damage to the settlement.

Sheep and goat bones also predominated in the assemblage from Late Bronze Age Termitito, but the relative frequency of caprines decreased to 56.2%, while cattle remains already represented 25.4% of the bones from domestic animals. The contribution of pig remains also increased, but this species still remained the third most exploited animal, yielding only 16.7% of the total of identifiable bones. A similar distribution of species was noted at the Early Bronze Age villages of Buccino in southern Italy (Barker in Holloway 1975: 60-61, Tables 3-4) and La Muculufa in Sicily. The frequency of sheep and goat bones was followed by pig and then by cattle remains in the assemblage found in the sanctuary of La Muculufa (Cruz-Uribe 1990: 52, Table 1) and during the whole Bronze Age at Broglio di Trebisacce. Cattle was the second best represented species at Coppa Nevigata (Bökönyi and Siracusano 1987) and in the transitional phase from the Late Bronze Age to the Early Iron Age at Broglio di Trebisacce (Tagliacozzo 1994b: 602, Table 2).

Hunting and fowling did not play an important role in the life of the peoples who inhabited the aforementioned Bronze Age sites, as is suggested by the small number of remains from wild species. The number and character of red deer remains from Termitito, however, are noteworthy. This species was exploited not only for meat and skin, but people took advantage of its hard tissue as well. Out of 37 remains, five antler and a metatarsus fragment showed marks of sawing and other forms of working. Because antler remains were found both attached to the frontal part of the skull as well as after separation, this kind of raw material was procured by killing the animals and also collecting shed antler (Chapter 5, Catalog No. 13-14). Red deer was also an important game at Broglio di Trebisacce during the Late Bronze Age, when its frequency increased to 59.4% in the assemblage. The rather uniform distribution of skeletal elements indicates that not only were antlers collected at the site, but also that animals' entire bodies were brought to the settlement (Tagliacozzo 1994b: 625, Table 25).

The remains from Termitito were greatly fragmented: only 12 bones (1.7%), including three long bones, were complete. Limb bones predominated over the bones from the head and trunk in most of the species, especially among caprines; proximal and middle segments of limbs—e.g., the humerus, femur, radius, and tibia, which hold most of the meat—were particularly frequent *(Table 3.3)*. This is also true for the remains from cattle, sheep, goat, and pig from Buccino (Barker in Holloway 1975: 60-61, Table 5). In contrast, skeletal parts from the head predominated among sheep and goat both at Broglio di Trebisacce (Tagliacozzo 1994b: 616, Table 20) and in the sanctuary and village of La Muculufa. The great number of isolated teeth is noteworthy at those sites. They point to an intensive post-depositional destruction of the assemblage, but also reflect the recovery methods employed, since one quarter of the sediment excavated at La Muculufa was sieved. It has also been observed

Skeletal part	Cattle	Sheep and goat	Pig	Ass	Dog	Ibex	Red deer	Roe deer	Wild boar	Wild cat	Hare
horn core/antler*							1				
neurocranium	13	12	10		1						
viscerocranium	1	1	1								
maxilla	2	7	5		3						
dentes	16	25	5						1		
mandibula	9	28	19		1		1				
Head Subtotal	**41**	**73**	**40**	**0**	**5**	**0**	**2**	**0**	**1**	**0**	**0**
atlas		3	1		1						
axis	2	1	1				1				
vert. cervicalis	8						1				
vert. thoracalis		2	1						1		
vert. lumbalis	2										
os sacrum											
pelvis	6	11	2	1	1						
vert. caudalis											
sternum											
costa	27	34	8				2		1		
Trunk Subtotal	**45**	**51**	**13**	**1**	**2**	**0**	**4**	**0**	**2**	**0**	**0**
scapula	3	11	8		1	1	1		1	1	
humerus	2	38	2								
femur	4	30	1								
patella	1										
radius	10	23	5				4				
ulna	7	4	5				1	1			
tibia	7	49	4								
fibula			2								
carpalia	3										
metacarpalia	5	13	3				3				
sesamoideum											
calcaneus	3	1	3								
astragalus	1	2	1								
centrotarsale							1				
metatarsalia	8	29	8		1		7				
ph. proximalis	1	1					1				
ph. media	4		1				3				
ph. distalis	3	2	1				1				
Limb Subtotal	**62**	**203**	**44**	**0**	**1**	**1**	**22**	**1**	**1**	**1**	**0**
Total	**148**	**327**	**97**	**1**	**9**	**1**	**28**	**1**	**4**	**1**	**1**

Table 3.3 Anatomical distribution of remains by species, Late Bronze Age Termitito. *Only antler attached to skull.

that young animals under 24 months and mostly the upper, meaty parts of the legs had been deposited in the sanctuary of La Muculufa (Cruz-Uribe 1990).

A conscious selection of meaty body parts both from domestic and wild species is also reflected by the distribution of remains representing different meat value categories at Termitito. Bones related to Category B meat were the most frequent in this assemblage, followed by Category A meat from caprines *(Fig. 3.2).* The strikingly small number of meatless distal limb bones in all species would indicate that they were left behind where they were slaughtered and underwent primary butchery. It is likely that mostly the meaty parts from the head, the shoulder- and pelvic girdle, as well as the proximal and middle segments of the limbs were taken to the settlement *(Table 3.3).*

The age distribution of remains at Termitito, like that at Late Neolithic Pantanello, showed that mostly adult animals were slaughtered. The remains of already developed individuals predominated even in pig *(Table 3.2).* This suggests that, in addition to the secondary exploitation of cattle and caprines, great attention must have been paid to preserving the stock sizes. At the Early Bronze Age site of Buccino, caprines and pig were exploited for meat purposes, while many cattle were reared to full maturity (Barker in Holloway 1975, Table 7). The age distribution of animals from Broglio di Trebisacce indicated that a large number (usually over 50%) of adult sheep, goat, and cattle were slaughtered. Aside from the ethnographically recorded use of pigs for sniffing out truffles, there is no other known secondary exploitation for swine, so the individuals from this species were mostly slaughtered before reaching the adult age; the latter age category was, however, also notably well-represented at Broglio di Trebisacce, yielding 21.8% of the pig remains (Tagliacozzo 1994b).

The exploitation of pig and dog showed particular characteristics at Termitito. As seen in Chapter 1 *(Fig. 1.31),* canine teeth with burnt and cracked tips were found in this assemblage. Relying on modern parallels, it has been suggested that singeing of pigs was already practiced and the canines would have suffered damage from this kind of removal of the coat (Takács 1990-1991). Out of nine dog remains, four (two maxillae and two mandibles) displayed traces of burning. Due to the small sample of dog bones and lacking other evidence of intentional (human) intervention, it is not possible to decide whether this species was occasionally eaten by people, or the affected remains came in contact with fire only accidentally. Although it seems unlikely that these dogs would have been sacrificed, it is relevant to note that dog remains were uncovered under the walls and floors of houses at the Bronze Age tell settlements of Tószeg–Laposhalom and Jászdózsa–Kápolnahalom in the Great Hungarian Plain. Skulls of other animals, such as aurochs, red deer, wild boar, and brown bear, were placed in a pit under the floor of the inner ditch at the latter settlement (Vörös 1996: 87).

In addition to the four dog remains, 30 burnt skeletal parts (altogether 4.3% of the assemblage) from cattle, sheep and goat, pig, and tortoise were identified. Fifteen skeletal parts, mostly tortoise carapaces, displayed cut marks. This is a notable increase compared to the two remains from the Neolithic assemblage, especially when taking into account that the earlier group was twice as large as the collection from Termitito. This diachronic increase in the incidence of cut marks must be related to the appearance of metal tools. The tendency is also paralleled by the small extent of bone manufacturing at this site. Only three worked bones were identified, two of them representing antler-debris (Chapter 5, Catalog No. 13-14).

The Archaic Greek Period

The bone remains at the 8th–6th century BC settlement at Incoronata were found in conjunction with accurately dated features that made phasing by century possible: (1) the 8th–7th century BC pre-colonial phase and (2) the mid-6th century BC colonial influx into the chora. Since the number of finds from the transition between the two periods was scarce, only the differences between the two centuries were considered in the analysis. Sheep and goat bones were most frequent in both periods by 38.9% and 46.1% respectively. During the 8th century, cattle was the second best represented species (31.1 %), closely followed by pig (27.9%). There was a change between the positions of those two species by the 6th century: contrary to the rather uniform representation of the listed species in the earlier period, caprines and pig predominated in the 6th century, while cattle yielded only 14.5% of the remains. Hunting did not play an important role in the life of inhabitants, as suggested by the small number of identified game species. The quantity of

Figure 3.4 Intact cattle metacarpus, Incoronata (IC 77-211P).

red deer remains is notable in the assemblage and they are equally represented in the two periods *(Table 3.2)*.

Parallels to our data come from the 9th–8th century BC and 6th–5th century BC levels of the indigenous site at Monte Maranfusa in Sicily. As was the case at Incoronata, sheep and goat were also the most frequently exploited animals at that site. Differences appear among cattle and pig, with pig the second most exploited species in the earlier assemblage, and cattle in the later. The representation of wild animals also showed a great similarity: a few species were hunted, but the number of red deer remains was notable in both groups (Di Rosa 2003).

Bone fragmentation at Incoronata was similar to that at Termitito, with only 1.6% of the remains found intact. Still, the number of long bones *(Fig. 3.4)* and horn cores is remarkable. The distribution of skeletal parts did not differ much between the two periods. Usually limb bones predominated, followed by the skeletal elements from head and trunk. Due to the great number of isolated teeth, the head remains from caprines and pig predominated in the assemblages dated to the 6th century *(Table 3.4)*. The same distribution in the assemblages from Monte Maranfusa did not show much difference between the two periods. Better representation of sheep and goat humeri in the later material was the only notable disparity that would have pointed to more consumption of good quality meat (Di Rosa 2003: 400-401, Tables 5-6).

In both periods at Monte Maranfusa, the grouping of skeletal elements according to Uerpmann's method indicates that bones connected to Category B meat, followed by the remains from Category A and Category C meat, were the most frequent. The same grouping of skeletal elements from Incoronata also indicated the predominance of remains from Category B meat in the earlier assemblage. These were followed, however, by the frequency of meatless bones (Category C), which, due to the great number of isolated teeth, ranked highest in the material deposited during the 6th century BC *(Fig. 3.2)*. This skewed distribution could be attributed to post-depositional damage rather than to changes in diet, but the accentuated presence of body parts with low economic value points to the fact that the processing of complete carcasses must have taken place at the settlement.

The age distribution of skeletal parts indicates that the domestic animals were mostly slaughtered as adults. A rather small proportion of remains originated from calves, lambs, and piglets in both periods. Almost 10% of caprine remains from the 6th century BC material belonged to senile animals, indicating their long term exploitation for milk, cheese, and probably wool *(Table 3.2)*. At Monte Maranfusa, the results indicate that young and developed animals were slaughtered equally in all species in both periods (Di Rosa 2003: 403-404).

The assemblage from Incoronata also yielded remains displaying damage caused by human intervention. Fifteen bones, mainly from meaty body parts, showed cut marks. The rib fragment of a pig displaying several parallel cut marks *(Fig. 3.5)* suggests the preparation of pork chops. Thirteen remains showed traces of burning. Only three artifacts were carved from bones, but one of them was especially important (Chapter 5, Catalog No. 15-17). The ground sheep astragalus found in a 6th century context indicated the use of these peculiar objects in certain religious or gaming activities. It has been supposed that the spread of gaming would have been connected to the Greeks, but religious practices related to astragali certainly date to even earlier times. The use of flattened and polished astragali from caprines has also been reported from the Late Bronze Age island site of Ustica I, north of Sicily, where five specimens were identified (Cruz-Uribe 1995: 92).

Figure 3.5 Rib fragment from pig displaying parallel cut marks, Incoronata (IC 78-92B).

Skeletal part	Cattle		Sheep and goat		Pig		Horse		Dog		Ibex		Red deer		Roe deer	Wild boar		Hare
	8th	6th	8th	6th	8th	6th	8th	6th	8th	6th	8th	6th	8th	6th	8th	8th	6th	8th
horn core/ antler*	5	2	6	7														
neurocranium	15	1	6	2	15	11				1							1	
viscerocranium	1	1	1															
maxilla	1	1	1		8	17	1		1							2		
dentes	18	24	40	196	20	176	1	1		1			2	2				
mandibula	14	13	28	20	28	55			1	1			4	0		2		
Head	54	42	82	225	71	259	2	1	2	3	0	0	5	2	0	4	1	0
atlas	3	2		1	1	1												
axis	1	2																
vert. cervicalis	8	5	1	1	1	0												
vert. thoracalis	1																	
vert. lumbalis		2	1	2				1										
os sacrum																		
pelvis	8	5	4	10	9	6				1								
vert. caudalis																		
sternum																		
costa	30	4	21	2	22	2							1					
Trunk	51	20	27	16	33	9	0	1	0	1	0	0	1	0	0	0	0	0
scapula	9	6	18	11	16	8	1				1							
humerus	3	6	22	21	10	6							3	2				1
femur	5	2	17	12	8	0			1				0	1				
patella	1																	
radius	2	6	21	18	2	0		1	1			1	1					
ulna	8	3	9	3	12	5	1							1				
tibia	16	3	41	29	14	2		1	1				2	3				
fibula					2	2												
carpalia	2	3																
metacarpalia	9	10	7	5	3	1									1			
sesamoideum																		
calcaneus	6	2	3	4	3	0			2							1		
astragalus	6	4		3	1	3			1				3	0				
centrotarsale	5			1									0	1				
metatarsalia	15	4	10	14	12	7							2	2				
ph. proximalis	6	6	2	4	1	2	2						5	2				
ph. media	6	3				1							1	1				
ph. distalis	7	0																
Limb Subtotal	106	58	150	125	84	37	4	2	6	0	1	1	17	13	1	1	0	1
Total	**211**	**115**	**259**	**366**	**188**	**305**	**6**	**4**	**10**	**4**	**1**	**1**	**23**	**15**	**1**	**5**	**1**	**1**

Table 3.4 Anatomical distribution of remains by species, Incoronata (8th–6th century BC). *Only antler attached to skull.

Skeletal part	Cattle	Sheep and goat	Pig	Horse	Ass	Dog	Cat	Aurochs	Ibex	Red deer	Fallow deer	Roe deer	Wild boar	Badger	Fox	Wolf	Hare
horn core/antler*	3	3								1							
neurocranium	29	2	3	1		2									1	1	
viscerocranium	2		1	1													
maxilla	2		2	1		1											
dentes	32	33	10	48	7	2				4					1		
mandibula	19	18	9	2		3		1		5				1		1	
Head	87	56	25	53	7	8	0	1	0	10	0	0	0	1	2	2	
atlas	7	2	3	1		3				1							
axis	6	1		2				1									
vert. cervicalis	12	1		12					1	1						1	
vert. thoracalis	10			3		1											
vert. lumbalis	18	2		6		2											
os sacrum						1											
pelvis	20	4	3	6		4				1							
vert. caudalis																	
sternum																	
costa	52	11		6		2				2							
Trunk	125	21	6	36	0	13	0	1	1	5	0	0	0	0	0	1	
scapula	29	5	2	5				1		4			3			1	
humerus	38	20	10	3		1		3		13		1					
femur	38	18	1	3		3		1		3	1						1
patella	1																
radius	25	16	3	2		1		1		11							
ulna	17	3	1	3						2			1		1		
tibia	43	30	6	11		2	1	2		8							
fibula						1											
carpalia	4			1													
metacarpalia	25	21	1	9						8	1		1			1	
sesamoideum																	
calcaneus	14	2		2				1		1							
astragalus	21	2	1	5						2		1					
centrotarsale	5			2						1							
metatarsalia	36	32	1	3		3				10	1				2	1	1
phalanx proximalis	32			7	1					6							
phalanx media	11	1	1	1						1	1						
phalanx distalis	2			2						1							
Limb	341	150	27	59	1	11	1	9	0	71	4	2	5	0	3	3	2
Total	553	227	58	148	8	32	1	11	1	86	4	2	5	1	5	6	2

Table 3.5 Anatomical distribution of remains by species, Pantanello Sanctuary (6th–3rd century BC). *Only antler attached to skull.

The Later Greek Period

The major change in animal husbandry in the chora of Metaponto began subsequent to the arrival of Greeks in the region. Two Pantanello sites evidence this change: the sanctuary (6th–3rd century BC) and the necropoleis (late 4th to early 3rd century BC). In contrast to the prehistoric period, when sheep and goat were the most preferred species, over half the bone remains identified from the Pantanello Sanctuary belonged to cattle *(Table 3.5)*. The presence of sheep and goat remains decreased to 21.4% of the total assemblage, while the third most important species was not pig, but horse (13.95%). The animals of the Hellenistic period (4th–1st century BC) included those in burial contexts (Bökönyi 1998). It is worth noting that ten times more caprine remains than other species were found in the Pantanello Necropoleis: four burials included sheep and goat bones; three included the combination of sheep/goat; cattle remains were found in the fill of two burials; and lastly, sheep/goat and pig remains were found in only one. Three burial pits (*fossa* tombs) included the more or less complete skeletons of a horse *(Fig. 3.6–7)*, a mule *(Fig 3.8)*, and a wolf *(Fig 3.9)*.

Figure 3.6 Horse burial, Tomb 316, Pantanello Necropoleis.

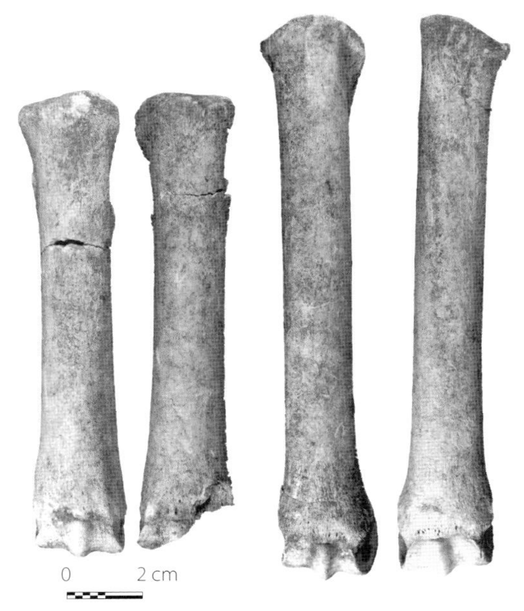

Figure 3.7 Horse metacarpi and metatarsi from Tomb 316, Pantanello Necropoleis (T 316).

A new species, the domestic cat, also made its first appearance at Metaponto *(Fig. 3.10)*. The dimensions of a tibia distal fragment from Pantanello were compared to the measurements of 49 recent and sexed specimens housed in the Osteological Collection of the Kiel University in order to appraise whether the bone belonged to a male or female. The ratio of the smallest diameter to the breadth of the distal epiphysis indicates that our specimen could have belonged to either a large female or smaller male cat *(Fig. 3.11)*. The measurements of two Roman period specimens, from Budapest–Albertfalva in the former province of Pannonia (Bökönyi 1974: 554) and from Quseir ael-Quadim in Egypt, ranked those finds among males. The latter, which belonged to a mummified cat, seemed to have been especially robust, a fact emphasized by the authors (Driesch and Boessneck 1983).

Changes in lifestyle during the Hellenistic period are also reflected by the greater number of wild mammals and a vulture, which were identified in remains from the Pantanello Sanctuary. Reminiscent of the prehistoric sites in the chora, bones of red deer also predominated among the wild species. These, together with the presence of the other nine mammalian species, offer evidence that people exploited both the woodlands and the open grassy fields to hunt for meat as well as animal skin and fur. Another Greek site in the northwest part of Lucania, the region of both an-

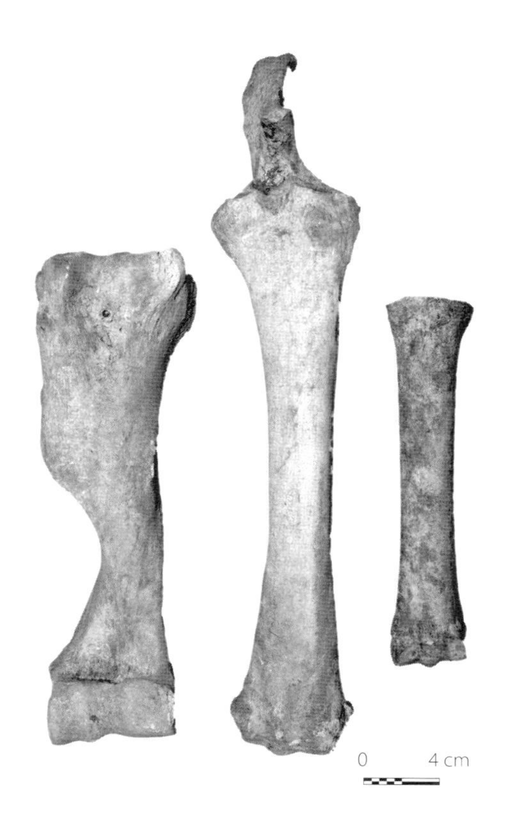

Figure 3.8 Mule humerus, radius, and ulna, and metacarpus from Tomb 62, Pantanello Necropoleis (T 62).

cient and modern Italy that includes Metaponto, is the fortified settlement at Roccagloriosa. This site yielded mostly sheep and goat remains, followed by cattle and pig bones. As at Pantanello, a greater number of wild animal species were identified, among which red deer was the best represented. Still, the number of remains at Pantanello attributed to identified wild species is low, making up only one fifth of the assemblage. (Bökönyi et al. 1993, Table 1).

As mentioned previously by Bökönyi in Chapter 1, the Pantanello Sanctuary yielded the second largest number of complete bones among all the sites, exceeded only by the later Roman period Kiln Deposit. Limb elements predominated in all species, indicating the transport and slaughtering of entire animals. Major meat-bearing bones, such as the humerus and femur, were well represented in cattle. Together with the skeletal elements of the trunk—the second best represented body region in this species—the evidence suggests a preference for good quality meat *(Table 3.5).* Indeed, the distribution of remains by the represented meat categories indicates that bones from the meaty parts of the body were almost as frequent as the most distal (meatless) skeletal elements in cattle *(Fig. 3.12).* Still, since the samples were hand-collected and not

Figure 3.9 Wolf burial, Tomb 321, Pantanello Necropoleis.

Figure 3.10 Distal fragment of cat tibia, Pantanello Sanctuary (PZ 91).

screened, the question remains as to whether cattle was indeed consumed differently, or if the distribution of skeletal elements illustrated here is due to the lack of smaller vertebrae from caprines and pig and the greater fragmentation of larger cattle bones.

The age distribution of remains indicates that mostly adult and mature specimens were consumed among cattle, sheep/goat, and pig alike. An interesting feature is the number of subadult specimens from horse, ass, and dog, indicating that half of these animals met their demise at a younger age, not typical for these species. Like the previous assemblages, however, only a small proportion of the bones could be aged because of their high degree of fragmentation. The ratio of bones of undetermined age is still well over 60% *(Table 3.2)*. The same analysis made on the material found at Roccagloriosa also showed that adult individuals were the most frequent in all domestic species, suggesting a strong interest in the secondary exploitation of animals (Bökönyi et al. 1993, Table 2).

Although the remains from the Pantanello Sanctuary showed that the main purpose of raising animals was to supply meat, only 10 remains displayed cut marks and 13 were burnt. A similarly small number of bones or antler was worked, only 0.5% of the whole assemblage.

The Roman Period

The assemblage found at the Pantanello Kiln Deposit (2nd century BC–1st century AD) was surprisingly similar to the material found at the Pantanello Sanctuary in terms of faunal composition and bone abundance. Cattle (45.2%), followed by caprines (27.8%), predominated in the assemblage. The third most frequent species was horse (9.9%), followed by dog (8.5%). Pig yielded only 7.6% of the remains. Red deer remained the most exploited species by 63.3% among wild animals *(Table 3.6)*.

At Sant'Angelo Grieco, a farmhouse site dating to the 1st century BC–1st century AD, the frequency sequence also showed cattle leading, followed by sheep and goat. Pig, horse and dog were scarcely represented in this assemblage. Red deer predominated among

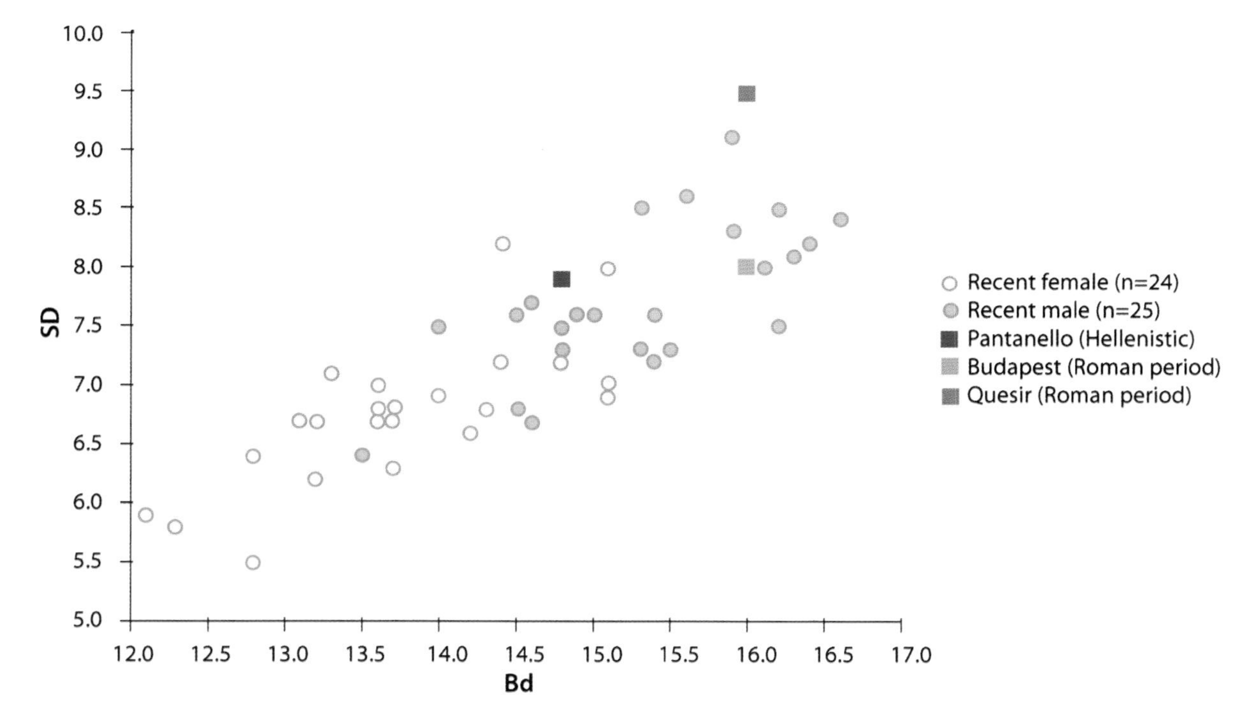

Figure 3.11 Ratio between the smallest diameter of the diaphysis (SD) to the breadth of the distal epiphysis (Bd) in cat tibia.

wild species. It is worth mentioning, however, that the number of animal bones found at Sant'Angelo Grieco was rather scanty, impeding its true comparison with other sites from the chora *(Table 3.7)*.

By contrast, at San Biagio, a 4th century AD villa, caprines predominated, although it must be noted that over 50% of these bones belonged to three full goat skeletons. Pig yielded a large number of bones to make up 29.5% of the assemblage. Interestingly, only 5.6% of the remains belonged to the cattle. Domestic hen furnished the same proportion of the material as cattle. The numbers of wild animal remains, as well as the game species, were negligible in this assemblage *(Table 3.8)*.

According to the frequency distribution of skeletal elements by body part, head bones ranked second—just after limb bones—among most species found in the Pantanello Kiln Deposit. This suggests the slaughtering and preparation of animals on location, as well as the discarding of food remains in the same place. The meaty leg bones were generally represented in the same proportion to the meatless metapodia. The number of isolated teeth was not great, but braincases and mandibles were relatively frequent, especially in cattle. These skeletal parts are related to Category B meat, which was best represented in cattle. The distribution of remains from sheep and goat did not show any differences.

At San Biagio, the greatest number of bones represented Category B meat, followed by Category A meat, for both caprines and pig *(Fig. 3.12)*. This proportion suggests that meatless body parts were sel-

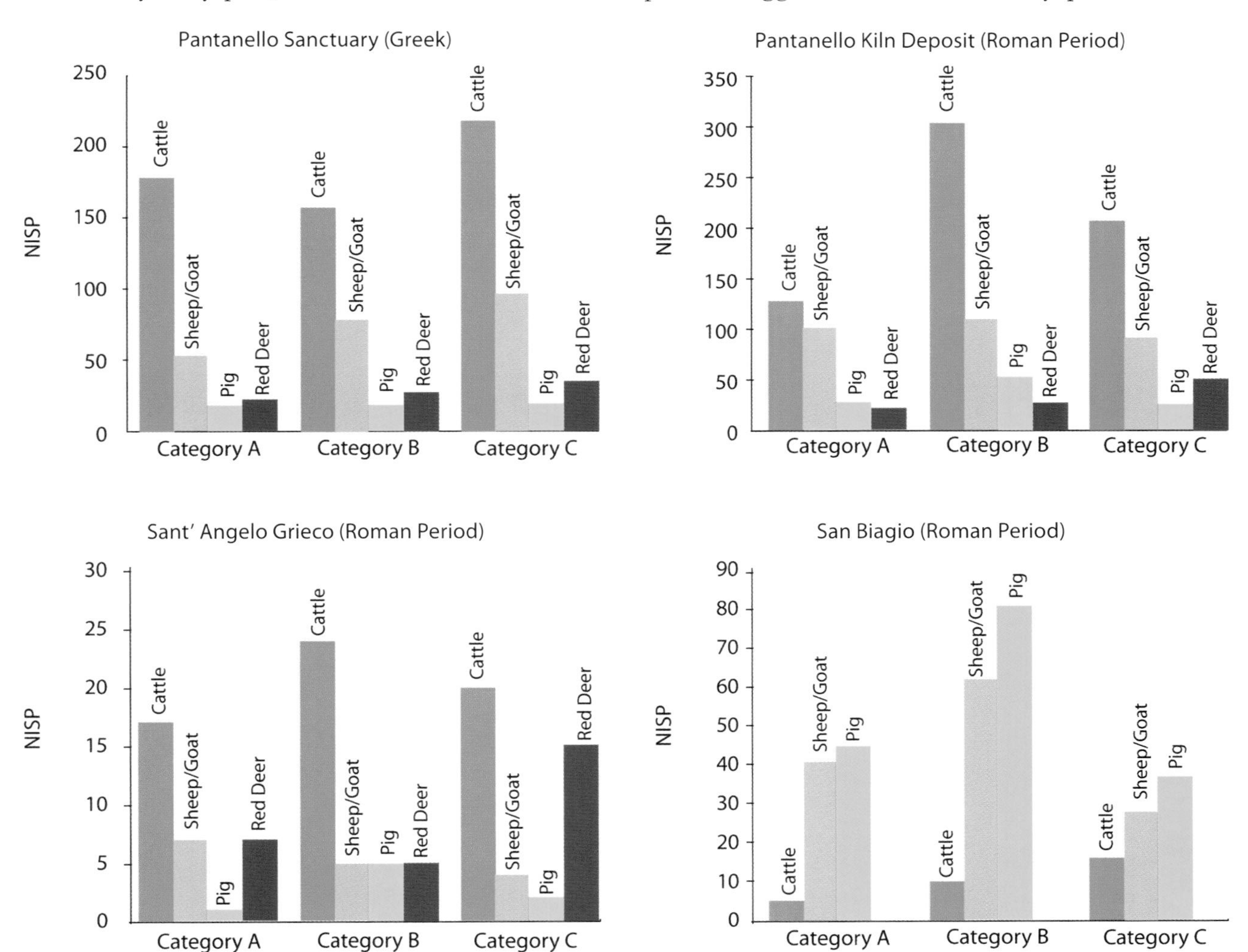

Figure 3.12 Distribution of remains by species and strata showing meat categories from the Pantanello Sanctuary, the Roman period Pantanello Kiln Deposit, the farmhouse at Sant' Angelo Grieco, and the villa at San Biagio. NISP = number of identified specimens. Category A = highest value meat; Category B = medium value meat; Category C = lowest value.

Skeletal part	Cattle	Sheep and goat	Pig	Horse	Ass	Dog	Aurochs	Ibex	Red deer	Roe deer	Wild boar	Fox
horn core/antler*	6	6										
neurocranium	86	8	4		1	3			1			
viscerocranium									1			
maxilla	4										1	
dentes	37	47	16	21	1	1			3		10	
mandibula	51	24	15	6	2	12	1		7			2
Head Subtotal	184	85	35	27	4	16	1		12	0	11	2
atlas	1	1				4					1	
axis	2					2	1		1			
vert. cervicalis	6	2		1		1						
vert. thoracalis	11	2		1								
vert. lumbalis	9					1			1			2
os sacrum	3			1								
pelvis	15	7	4	2		4			3		1	1
vert. caudalis												
sternum												
costa	84	34	12	1		25			2			1
Trunk Subtotal	131	46	16	6		37	1		7	0	2	4
scapula	20	13	9	4		1	3		5		1	
humerus	27	31	14	8		8	1	1	9	1	2	1
femur	35	47	1	4		8			2			1
patella	2											
radius	31	45	4	9		8	1		5	1		1
ulna	20	5	5	5		6	2		5	1		
tibia	30	84	9	16		9			7		2	
fibula			5	6		1					1	
carpalia	32	1		5								
metacarpalia	30	19	2	8	1	3	1		13			
sesamoideum												
calcaneus	16	2	3	6		3			6		1	
astragalus	14	1		4	1				3			
centrotarsale	8	1		9								
metatarsalia	39	14	3	5		17	1		21			2
phalanx proximalis	9	1	2	11		4			2			
phalanx media	6			4					2			
phalanx distalis	7			3								
Limb Subtotal	326	264	57	107	2	68	9		80	3	7	5
Total	**641**	**395**	**108**	**140**	**6**	**121**	**11**	**1**	**99**	**3**	**20**	**11**

Table 3.6 Distribution of remains by species, Pantanello Kiln Deposit (2nd c. BC–1st c. AD). *Only antler attached to skull.

Skeletal part	Cattle	Sheep and goat	Pig	Horse	Dog	Red deer	Wild boar
horn core/antler*		1				11	
neurocranium			1			1	
viscerocranium							
maxilla							
dentes	3		2		1	2	
mandibula	6	1	1				
Head Subtotal	9	2	4	0	1	14	0
atlas			1				
axis	1						
vert. cervicalis	2						
vert. thoracalis						1	1
vert. lumbalis						3	
os sacrum							
pelvis							
vert. caudalis							
sternum							
costa	9	3		2			
Trunk Subtotal	12	3	1	2	0	4	1
scapula	2						
humerus	1	3		1			
femur	2	1			1	3	
patella	1						
radius	2	2		1		1	
ulna	2		1	1			1
tibia	13	1	2			3	
fibula							
carpalia	2						
metacarpalia	3						
sesamoideum							
calcaneus							
astragalus	6						
centrotarsale							
metatarsalia	3	2		1		1	
phalanx proximalis	3	1				1	1
phalanx media							
phalanx distalis							
Limb Subtotal	40	10	3	4	1	9	2
Total	**61**	**15**	**8**	**6**	**2**	**27**	**3**

Table 3.7 Distribution of remains by species, Roman farmhouse, Sant' Angelo Grieco (2nd c. BC–1st c. AD). *Only antler attached to skull.

Skeletal part	Cattle	Sheep and goat	Pig	Horse	Dog	Hen	Red deer	Wild boar	Hare
horn core/antler*									
neurocranium	1	2	22						
viscerocranium		2							
maxilla			1						
dentes	5	12	22	1					
mandibula	2	9	17	1					1
Head Subtotal	8	25	62	2	0	0	0	0	1
atlas									
axis		2							
vert. cervicalis			2						
vert. thoracalis	1	1							
vert. lumbalis									
os sacrum									
pelvis	1	5	11						
vert. caudalis									
sternum									
costa	2	18	11				1		
Trunk Subtotal	4	26	24	0	0	0	1	0	0
scapula	2	9	14			7			
humerus		15	8		1	7		1	
femur	1	9	10			6	1		2
patella									
radius		13	7			1		1	
ulna	1	5	10			1			
tibia	4	15	13			6	1		
fibula			1			1			
carpalia									
metacarpalia		3							
sesamoideum									
calcaneus	1	3	2		1				
astragalus	1	2	1						
centrotarsale	2								
metatarsalia	1	2	9	1					
phalanx proximalis	1	4	2	1					
phalanx media	4								
phalanx distalis	1								
Limb Subtotal	19	80	77	2	2	29	2	2	2
Total	**31**	**125**	**163**	**4**	**2**	**29**	**3**	**2**	**3**

Table 3.8 Anatomical distribution of remains by species, Roman period villa, San Biagio (3rd–4th century AD) *Only antler attached to skull..

dom taken to the settlement *(Table 3.8)*. The ageing of remains indicated that fully developed specimens were almost exclusively slaughtered among cattle and caprines. Piglets, however, were better represented than adult pigs, for the second time in the periods of the chora that have been studies. Following the logic of secondary exploitation, it is likely that hens were kept for egg production rather than for meat as only the bones of adult hens were identified.

A preference for young pig meat may already be seen by the early phase of the Roman Empire in the distribution of remains at the Pantanello Kiln Deposit. Meat from cattle, sheep, and goat was provided by slaughtering adult and mature individuals. The secondary exploitation of these animals as draught cattle, as well as for milk and wool production, was well-established during the entire Roman period, similarly to the previous periods in the region *(Table 3.2)*.

In spite of the fact that the identified remains represented food refuse, only 10 bones displayed cut marks. More remains (41) showed traces of burning in this assemblage. Interestingly, 15 of these finds belonged to dogs. Although there is evidence for dog consumption during the Roman Period, as Bökönyi mentioned in Chapter 1, dogs were largely appreciated and bred by the Romans in Pannonia. Since butchery marks did not evidence their consumption in Pantanello, these remains may have accidentally come in contact with fire.

Bone manufacture was relatively advanced during this age, producing not only a number of tools used in daily life, but fine objects as well. The most unusual artifacts were found in the Pantanello Kiln Deposit, but the use of bone artefacts survived as late as the end of the Roman Period as shown by the few sophisticated specimens found at San Biagio (see Chapter 5).

Conclusions

The detailed analyses of the archaeozoological assemblages found in the chora of Metaponto indicate a uniform tendency in animal exploitation from the Neolithic through the early Archaic periods. Sheep and goat, followed by the much smaller representation of cattle and pig, were the most common animals in the assemblages of Late Neolithic Pantanello, Late Bronze Age Termitito, and even in the 7th century BC deposit of Incoronata. Most of the ruminants were exploited for secondary products such as milk, easily turned into cheese for storage and trade. Adult sheep, shorn for several seasons, provided more wool, a valuable commodity. Training cattle to work a plough or pull a wagon also took time. Longevity for these species is indicated in the age distribution of finds. Pigs were also often reared to maturity, but show a less consistent age pattern. The distribution of bones by body parts indicates that most animals were transported to the site as entire carcasses at both Pantanello and Incoronata. The Late Bronze Age site of Termitito represents an exception, since the strikingly small number of remains from the distal limbs indicated that the meatless skeletal elements must have been left behind, and only the meaty body parts were taken to the settlement. This suggests that consumption rather than processing took place at the site.

At 6th century BC Incoronata the increasing number of pig remains, as well as the presence of hen, already suggest a shift to a more sedentary way of life. By the time of the Greek settlement at Pantanello, meat provision was mostly based on cattle husbandry. As opposed to the domestic consumption of small-bodied sheep or goat by an extended family, slaughtering these large animals profitably may have required a more centralized distribution of beef at the colony. Horse became the third best represented species. In addition to hen, the domestic cat as well as the domestic ass and mule are found. Changes in lifestyle in the Hellenistic period are also evidenced by the increasing number and character of hunted species.

This general pattern of animal husbandry and hunting continues into the Roman period at Pantanello. As in the earlier age, the distribution of skeletal elements suggests that animals were slaughtered and prepared at the site. The only change in animal consumption seems to have been the increased killing of piglets. This tendency is seen at San Biagio at the end of the Roman period, where pork seems to have been the preferred meat (the numerical predominance of caprine bones in the assemblage, as noted, is due to the three full goat skeletons). Hens were also greatly appreciated, both for meat and eggs. According to the species composition and distribution of skeletal elements by body parts and age profiles of animals, it is likely that a better stocked pantry and more abundant table was maintained in the villa at San Biagio during the late 4th century AD.

4
Bird Remains from the Chora of Metaponto

Erika Gál

Introduction

The impressive archaeozoological assemblage from the Metaponto chora yielded approximately 12,000 remains. Of the 7,000 bones which could be identified, 61 were determined to be avian. The majority of bird remains (27 bones, equal to 4.7% of the total bone assemblage) came from San Biagio, a villa inhabited during the late 3rd–4th century AD. The other sites yielded only 1 to 15 remains each *(Table 4.1)*.

There may be several reasons for the under-representation of bird remains in these hand-collected assemblages. First, the economic value of birds, especially in prehistoric societies when only wild birds were available, is negligible in comparison with larger mammals that could be exploited. Second, small and fragile bird bones—especially the porous skeletal parts of young individuals—are more vulnerable to a number of taphonomic agents both before and after deposition. Dogs, cats, and pigs, for example, are potential consumers of small and soft bones. Finally, the lack of fine recovery techniques, such as water sieving or dry screening during excavation, tends to exacerbate the under-representation of small finds.

The archaeological periods represented by the avian bones range from the Neolithic pits at Pantanello to the late Roman site at San Biagio. Given the chronological, palaeoecological, and cultural differences among the sites under study, a review of the identified bones and species, as well the importance of fowling and/or poultry keeping, is given in the case of each site in chronological sequence.

Neolithic Pantanello

The Neolithic pits at Pantanello yielded 9 avian remains. These were identified as having belonged to mallard (*Anas platyrhynchos* Linnaeus, 1758), partridge (*Perdix perdix* Linnaeus, 1758), and rook (*Corvus frugilegus* Linnaeus, 1758). The mallard remains, indicating two individuals, included a distal fragment of femur from the left leg, a proximal and a distal fragment of tibiotarsus from the left leg, a fragmented tarsometatarsus, as well as a phalanx. The bones from the partridge—a coracoideum and a scapula fragment from the right shoulder—most likely belonged to the same individual *(Table 4.2)*. The only rook bone is a proximal part of an ulna from the right wing. The

Site	Period	Bird NISP	Total Vertebrate NISP	Percentage of Total NISP
Pantanello	Late Neolithic	9	1360	0.66%
Termitito	Late Bronze Age	1	692	0.14%
Incoronata	8th–6th century BC	1	1668	0.06%
Pantanello Sanctuary	6th–3rd century BC	12	1186	1.01%
Pantanello Kiln Deposit	2nd century BC–1st century AD	10	1598	0.62%
Sant'Angelo Grieco	6th century BC–1st century AD (primarily 2nd c. BC–1st c. AD)	1	125	0.80%
San Biagio	Late 3rd–4th century AD	27	568	4.75%

Table 4.1. Number, distribution, and percentage of identified avian remains from the archaeozoological assemblages, Metaponto chora. (NISP = number of identified specimens.)

Species	Site and Period													
	Pantanello Neolithic		**Termitito** Bronze Age		**Incoronata** 8^{th}c. BC– 6th c. BC		**Pantanello Sanctuary** 6th c. BC– 3rd c. BC		**Pantanello Kiln** 2nd c. BC– 1st c. AD		**Sant'Angelo Grieco** 1st c. BC– 1st c. AD		**San Biagio** 4th c. AD	
	NISP	MNI	NISP	MNI	NISP	MNI	NISP	MNI	NISP	MNI	NISP	MNI	NISP	MNI
Anas platyrhynchos Mallard	5	1												
Aegypius monachus Black vulture							8	1						
Perdix perdix Partridge	2	1												
Gallus domesticus Domestic hen					1	1	2	1	10	2	1	1	27	4
Corvus frugilegus Rook	1	1												
Corvus corone/cornix Carrion/hooded crow							1	1						
Aves sp. indet. unidentifiable	1		1	1			1	1						
Total	9	3	1	1	1	1	12	4	10	2	1	1	27	4

Table 4.2 . Species, number of identifiable specimens (NISP), and calculated minimal number of individuals (MNI) by site.

measurements from all bird bones included in this study are summarized in Table 4.3.

Although only a few species were identified from this site, they represent various environmental conditions. Mallard is an aquatic fowl living in a wide range of habitats, from flowing and stationary fresh waters through brackish waters to shallow coastlines. This omnivorous species, characterized by a variety of diet and feeding habits, usually nests on the ground or in hollow trees. Partridge, by contrast, prefers contiguous grasslands and avoids swamps, forests, and the coastline. This species, which feeds mainly on plant materials and occasionally on insects, nests on the ground in thick vegetation. The rook, in turn, is an arboreal species that requires fairly tall trees for breeding. It lives on the edges of forests while avoiding dense woodland and wetlands. An omnivorous bird, the rook eats invertebrates (mainly beetles and earthworms), plant material (principally cereal grain), small vertebrates, carrion, and scraps of all kinds. Outside the breeding season, the rook is gregarious when feeding, roosting and migrating. Rooks have recently become considered birds of agricultural landscapes (Cramp 1998), and opinions differ on their role: they may be considered useful in an economic sense, since they kill pests, but farmers often persecute them as a vermin feeding on newly sowed seeds in cultivated land.

Bronze Age Termitito

The Bronze Age site of Termitito yielded only a single avian bone. The diaphysis fragment most probably belonged to a humerus, and the bird species could not be determined due to the level of fragmentation. According to the size of the bone fragment, however, it possibly belonged to a goose-sized bird.

Iron Age Incoronata

The site of Incoronata also yielded a single bird bone, this time from a 6th century BC feature. Nevertheless, this remain is one of the most significant avian bones in the material, since the diaphysis fragment of a tibi-

Species	Skeletal part	Side	Site	Inventory No.	Age	GL	SL	Bp*	Dp**	SD	Bd***	Dd****
Anas platyrhynchos	femur	sin	Pantanello Pits	PZ Pit I	Neolithic						12.0	8.3
Anas platyrhynchos	tibiotarsus	sin	Pantanello Pits	PZ Pit I	Neolithic			13.9				
Anas platyrhynchos	tibiotarsus	sin	Pantanello Pitso	PZ Pit I	Neolithic					4.1	9.4	10.1
Anas platyrhynchos	tarso-metatarsus	sin	Pantanello Pits	PZ Pit I	Neolithic			9.7	9.4			
Aegypius monachus	humerus	sin	Pantanello Sanctuary	PZ-78-359B	6th–3rd c. BC			53.0				
Aegypius monachus	femur	sin	Pantanello Sanctuary	PZ-78-426B	6th–3rd c. BC		125.0	37.1	21.1	15.1	36.4	26.0
Aegypius monachus	phalanx 1 digiti 2	sin	Pantanello Sanctuary	PZ-78-359B	6th–3rd c. BC	38.4		12.4	12.5	7.8	9.3	9.9
Aegypius monachus	tarso-metatarsus	sin	Pantanello Sanctuary	PZ-78-359B	6th–3rd c. BC			26.5	20.4			
Aegypius monachus	phalanx pedis 3 digiti 2	sin	Pantanello Sanct.	PZ-78-432B	6th–3rd c. BC	35.0						
Perdix perdix	scapula	dex	Pantanello Pits	PZ 83 Pit JJ	Neolithic			8.2		2.5		
Gallus domesticus	coracoideum	sin	San Biagio	SB-80-359	3rd–4th c. AD	48.7				5.1	13.4	11.8
Gallus domesticus	coracoideum	dex	San Biagio	SB-80-379B	3rd–4th c. AD	52.3				4.8		10.3
Gallus domesticus	coracoideum	sin	San Biagio	SB-80-384B	3rd–4th c. AD	54.1				5.0		11.1
Gallus domesticus	coracoideum	dex	San Biagio	SB-80-379B	3rd–4th c. AD	55.7				4.7	14.0	11.4
Gallus domesticus	coracoideum	sin	San Biagio	SB-80-240B	3rd–4th c. AD	58.9				5.7		13.0
Gallus domesticus	humerus	dex	San Biagio	SB-80-324B	3rd–4th c. AD	ap. 76.0				7.9		
Gallus domesticus	humerus	sin	San Biagio	SB-80-320B	3rd–4th c. AD	76.2		20.3		7.7	15.7	8.9
Gallus domesticus	humerus	sin	San Biagio	SB-80-372B	3rd–4th c. AD					6.7	13.3	7.5
Gallus domesticus	femur	dex	Pantanello Kiln	PZ-76-277B	2nd c. BC–1st AD	69.6	66.2	13.4	9.5	5.9	13.0	10.8
Gallus domesticus	femur	dex	Pantanello Kiln	PZ-75-525	2nd c. BC–1st AD	71.4	67.8	13.0	9.2	6.2	13.4	11.3
Gallus domesticus	femur	dex	San Biagio	SB-80-372B	3rd–4th c. AD	72.7		14.5	9.7	5.8	13.4	11.4
Gallus domesticus	femur	sin	Pantanello Kiln	PZ-77-834P	2nd c. BC–1st AD	77.2		15.3	10.7	6.4	14.9	12.4
Gallus domesticus	femur	sin	San Biagio		3rd–4th c. AD	80.0		16.4	11.7	7.3	13.4	
Gallus domesticus	femur	sin	Pantanello Kiln	PZ-81-378B	2nd c. BC–1st AD			15.0	9.3			
Gallus domesticus	carpometa carpus	sin	Pantanello Kiln	PZ-81-536	2nd c. BC–1st AD	42.1		12.8		10.3		
Gallus domesticus	tibiotarsus	sin	Pantanello Kiln	PZ-77-834P	2nd c. BC–1st AD	106.6				5.7	10.7	11.2
Gallus domesticus	tibiotarsus	dex	Pantanello Kiln	PZ-76-277B	2nd c. BC–1st AD	107.4			18.0	5.6	10.6	11.3
Gallus domesticus	tibiotarsus	sin	Pantanello Kiln	PZ-76-277B	2nd c. BC–1st AD				16.6	5.4		
Gallus domesticus	tibiotarsus	dex	San Biagio	SB-80-384B	3rd–4th c. AD				21.0	7.5		
Gallus domesticus	tibia	dex	San Biagio	SB-80-320B	3rd–4th c. AD						10.3	
Gallus domesticus	tarso-metatarsus (female)	dex	Pantanello Kiln	PZ-81-120B-G	2nd c. BC–1st AD	67.8		11.5	10.4	6.1	11.6	9.4
Corvus corone/cornix	ulna	sin	Pantanello Sanctuary	PZ-75-867x	2nd c. BC–1st AD	83.6		10.1	11.8	4.8	10.1	11.6

Table 4.3 Sizes of avian skeletal parts in millimeters, following von den Driesch1976 (*Dic in scapula and Dip in tibiotarsus, **Dip in ulna, ***Bb and ****BF in coracoideum).

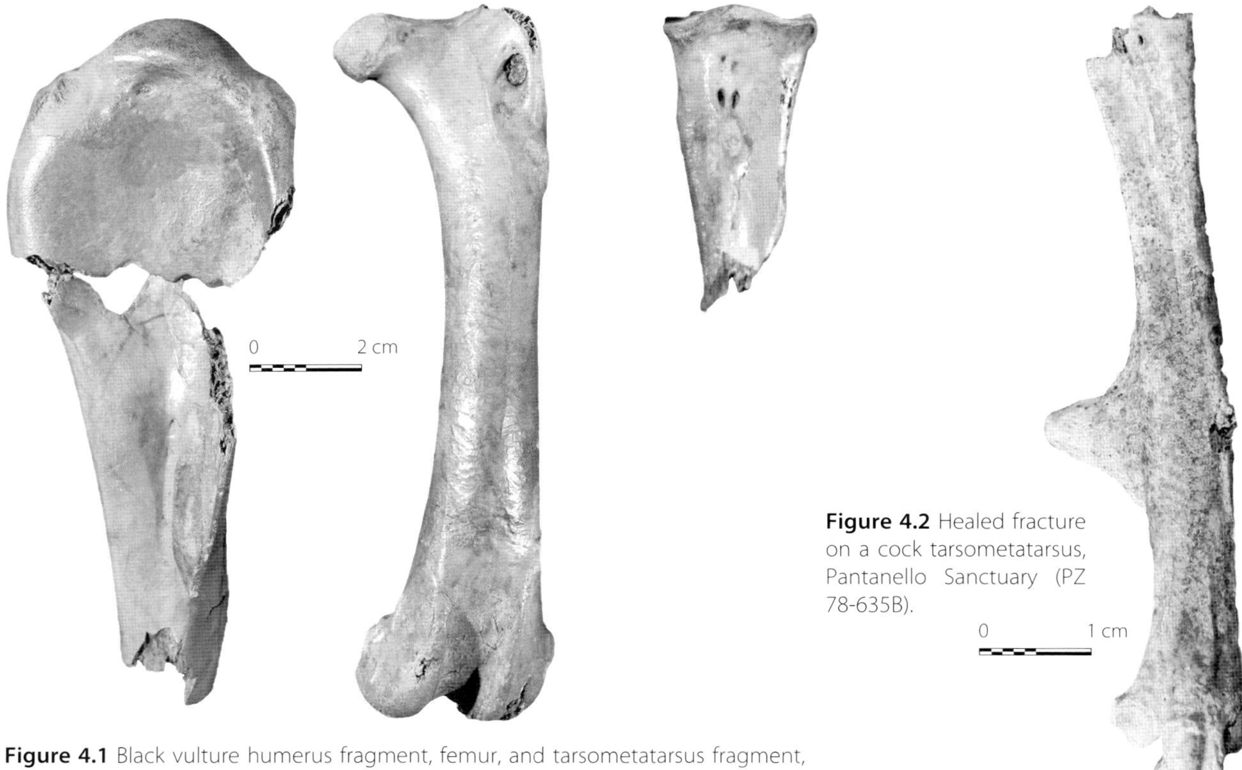

Figure 4.2 Healed fracture on a cock tarsometatarsus, Pantanello Sanctuary (PZ 78-635B).

Figure 4.1 Black vulture humerus fragment, femur, and tarsometatarsus fragment, Pantanello Sanctuary (PZ 78-359B).

otarsus was identified as having belonged to the domestic hen (*Gallus domesticus* Linnaeus, 1758). Owing to the poor preservation of the remains, the size of the hen cannot be estimated and compared with the sizes of other early specimens from Europe.

The Greek Period Sanctuary at Pantanello

The Pantanello Sanctuary, where animal bones were deposited between the 6th and 3rd centuries BC, yielded 10 avian remains, 8 of which belonged to a black vulture (*Aegypius monachus* Linnaeus, 1766). These bones, representing an incomplete skeleton, included a fragment from the mandible, proximal parts of the left humerus and tibiotarsus, a complete left femur *(Fig. 4.1)*, an ulnare, a fragment of the pelvis, a phalanx from the foot, and a talon from the second toe. Each skeletal part had a faint brown color. Since the coloring was light and uniform, it was ascribed to the deposition of the skeleton in anaerobic conditions, rather than to burning. The black vulture is one of the largest diurnal birds of prey, a solitary forager over large open terrains—either uninhabited mountains or lowland plains—where there are stocks of large grazing animals. It rarely takes live prey, but feeds on carrion, mainly from carcasses of middle to large size animals. It usually nests in high trees, but may build on cliff ledges even when trees are available.

One bone, a complete left ulna, was identified as that of carrion/hooded crow (*Corvus corone/cornix* Linnaeus, 1758). Distinguishing between a rook and a crow is difficult because of the morphological similarities between the two species and the overlap in bone size. Using the mean sizes of the ulna from 68 rooks and 70 crows (Tomek and Bocheński 2000, 42-43, Table 10), the specimen from Pantanello was identified as a crow. Crows are raucus arboreal song birds, living in groves or at the edges of forests. Like rooks, they avoid dense woodlands. An omnivorous bird, the crow eats a wide range of plant materials, invertebrates, small vertebrates, and even carrion, foraging almost exclusively on the ground. Crows live in large and clamorous flocks (Cramp 1998), nesting in the crowns of high trees and almost always in isolated groups.

Two bones from this site were classified as domestic hen. In addition to a humerus diaphysis fragment, a distal tarsometatarsus fragment belonging to a cock was found. Evidence for a healed fracture above the spur could be seen on the latter specimen *(Fig. 4.2).*

Roman Period Pantanello

Bones from domestic hens were found in the Pantanello Kiln Deposit, dated to the 2nd century BC–1st century AD. These consisted of seven complete bones and three fragments, and represent at least two individuals. One of the tarsometatarsi showed no sign of a spur and therefore must have belonged to a female.

Roman Period Sant'Angelo Grieco

Inhabited from the 1st century BC–1st century AD, the farmhouse site at Sant'Angelo Grieco yielded only a tibiotarsus fragment from a domestic hen.

Roman San Biagio

As mentioned earlier, most of the avian remains (27 bones, equal to 4.7% of the total bone assemblage) came from the late 3rd–4th century Roman site at San Biagio, the latest of the studied settlements in the chora of Metaponto. According to the best represented skeletal part—four left side and three right side coracoidei—the bones, belonging to domestic hens, came from at least four individuals. Although the number of studied remains is insufficient for statistical evaluation, it is noteworthy that the meaty body parts such as the coracoideum and humerus from the shoulder girdle and wing, as well as the femur and tibiotarsus from the leg, were the best represented *(Fig. 4.3).* The ulna, radius, and fibula yielded one remain each. This small number of distal wing bones may be related to the poor economic value and removal of this body part during processing. On the other hand, the slim and fragile bird radius, along with the fibula, is often missing from archaeozoological bone assemblages.

Discussion

In comparison with the mammalian bones, avian remains were underrepresented in the bone assemblages found in the chora of Metaponto. Except for the 6th–3rd century Sanctuary at Pantanello and the 4th century AD settlement at San Biagio, bird bones made up less than 1% of all animal bone remains. Birds may have had only a marginal role even in the economies of those two sites, since they represented 1.01% and 4.75% of bone assemblages respectively *(Table 4.1).*

According to a recently published review of animal exploitation in central and south Italy, bird remains were found at eleven Early Neolithic sites. With the exception of Grotta dei Piccioni, where avian remains made up more than half of the otherwise rather small bone sample (the total NISP—number of identifiable specimens—was 255), bird bones usually were underrepresented (less than 10%) in the assemblages (Tagliacozzo 2005-2006, 436, Table 5). It should be mentioned that four sites of the eleven were caves, where not only people, but also birds of prey and mammalian predators may have contributed to the deposition of avian remains. There are a number of owl species, but also some diurnal birds of prey, which nest or roost in the vicinity of cave entrances. The undigested food remains emitted in the form of pellets may largely increase the content of fossil deposits. Moreover, water and wind transport through the fissures of caves may also contribute to bone deposition from species which otherwise do not have any connection with cave environments (Gál, 2008c). A number of prehistoric bone assemblages from cave and open-air sites are known from southern Europe, but since most of the avian remains have not been published on the level of species identification, little is known of the prehistoric distribution of the species under discussion here.

Mallard was the most preferred game bird by 44.7% of the avian remains for people inhabiting Grotta della Madonna in Calabria during the Upper Palaeolithic (Tagliacozzo and Gala 2002, 119, Table 1). It was also caught, but with much less frequency, at Grotta Romanelli in Apulia (Cassoli and Tagliacozzo 1997, 305, Table 1). Mallard was identified at the Early Neolithic village La Marmotta (Tagliacozzo

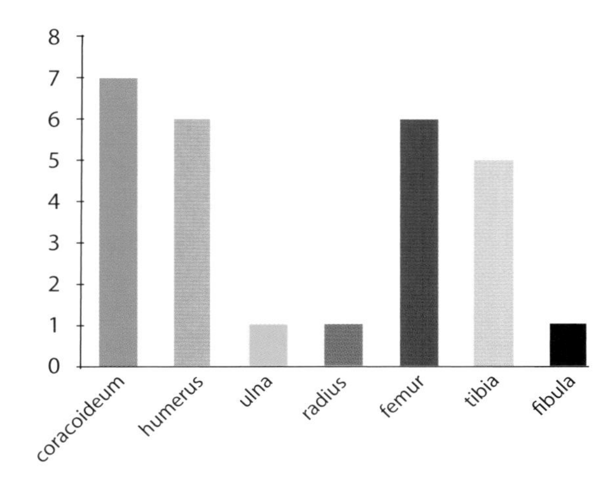

Figure 4.3 Distribution of skeletal parts (n=27) from domestic hen, 3rd–4th century AD villa, San Biagio.

2005-2006, 433) and another Neolithic site, Grotta Continenza (Wilkens 1989-1990, 93) in Central Italy; remains were also found in the Late Bronze Age layers of Coppa Nevigata in Puglia (Bökönyi – Siracusano 1987, 206, Table 1).

Ducks must have been among the favorite wild fowl hunted during prehistoric times in Italy. In addition to the mallard, ferruginous duck (*Aythya nyroca* Güldenstadt, 1770), shoveler (*Anas clypeata* Linnaeus, 1758), and pochard (*Aythya ferina* Linnaeus, 1758) were reported from La Marmotta, and a wild duck (*Anas* sp.) from the Early Neolithic ("Impressed Ware") site of Rendina in the region of Basilicata (Bökönyi 1985, 184, Table 1). Mallard and quail were also found in the Neolithic layers of Sitagroi in Northeast Greece (Bökönyi 1986, 69, Table 5.2b).

The only reported data found for partridge comes from the Early Neolithic levels of Grotta Continenza (Wilkens 1989-1990, 93). Hooded crow (*Corvus corone cornix* Linnaeus, 1758), the closest relative of the rook, was identified at the Early Neolithic site of La Marmotta and in the Middle Bronze Age layers of Broglio di Trebisacce in the region of Calabria (Tagliacozzo 1994, 602, Table 9).

Mallard, partridge, rook, and crow live year-round in Italy. Mallard and partridge have traditionally been preferred game, sought after for their taste. Crows are considered inedible according to modern tastes, but nevertheless were consumed by the 17th century Hungarian aristocracy (Benda 2004, 223). According to the ethnographic record, crows were trapped by noose and prepared with fried eggs as late as the beginning of the 20th century (Kardos 1943, 11).

Remains of black vultures, like those from the Pantanello Sanctuary, are unusual and rare finds in archaeological deposits. In Italy, this species had already been identified from the Upper Palaeolithic site of Grotta Romanelli. Out of the 31,984 avian remains recovered, 10 came from this large diurnal bird of prey. Like the Pantanello finds, those remains did not display any signs of human intervention (Cassoli and Tagliacozzo 1997, 307, Table 1). The distal fragment of a humerus from a vulture was reported from the Bronze Age site of La Starza in Southern Italy. Owing to the partial preservation of that skeletal part, it could not be definitely determined whether it belonged to a griffon vulture (*Gyps fulvus* Hablizl, 1783) or a black vulture (Albarella, 1997). Black vulture was also identified from two remains at the 4th to 1st century BC deposit of Lattara in France, an Iron Age town founded in the 6th century BC (Garcia Petit 2005).

The most interesting black vulture finds in the Mediterranean region, however, are represented by the number of flutes found in Spain and Portugal. Evidence from the Roman period up to recent times indicates the use of ulnae—from both black and griffon vultures—as raw material for producing wind instruments in the Iberian Peninsula (Moreno-García et al. 2005). Manufacturing flutes from the skeletal part of vulture was not only a custom in Europe, but also in the Near East. Evidence for a vulture flute comes from the Great Mosque of ar-Rafiga in Western Syria, dated to the middle of the 11th century AD (Becker 2005).

Mentioning these special finds seems relevant here since, as Becker points out, a particular relationship has existed between people and vultures for a long time. The special role of these large raptorial birds is evinced both by artistic representations (such as wall paintings, stone figurines, and reliefs) and bone evidence in the ancient Near East and in the later Greek and Roman cultures (Becker 2005, 334). The earliest black vulture flute in Spain also came from a ritual deposit. It was found in a funerary context, next to the pantheon of a hispano-visigothic family. It has been noted that vultures were associated with death and it was believed in Iberian pre-Roman times that they carried off the soul of the deceased (Moreno-García et al. 2005, 332-333).

Although traces of human intervention were not detected on the skeletal parts of the black vulture found in the Pantanello Sanctuary, and the bone artifact collection from this site did not include items made from avian bones, we may not exclude the possibility of using ulnae for this purpose. The partial skeleton included bones from all body parts, but was missing both ulnae. Moreover, two of the smallest bones recovered from the wing (an os carpi ulnare and a phalanx 1 digiti 2) would suggest that the wing was not used in its entire size but perhaps dismembered.

The exploitation of large feathers has been postulated as the main purpose for killing the vulture (along with a crane) at the Bronze Age site of La Starza (Albarella 1997, 348). Feathers have been widely used in the cultural and spiritual life of people. They could represent social signalling, display, and decoration

(Serjeantson 1997, 257). One tantalizing explanation for exploiting vulture plumage can be found in Pliny, who offers the following advice:

> Among the birds that afford us remedies against serpents, it is the vulture that occupies the highest rank; the black vulture, it has been remarked, being less efficacious than the others. The smell of their feathers, burnt, will repel serpents, they say; and it has been asserted that persons who carry the heart of this bird about them will be safe, not only from serpents, but from wild beasts as well, and will have nothing to fear from the attacks of robbers or from the wrath of kings.[1]

The uniform color of bone finds and even representation of body parts in the case of the Pantanello vulture would indicate the separation and use of a section of the wing or only the feathers, and the discarding of the rest of the carcass. This is also suggested by the lack of radii, ulnae, and carpometacarpi, skeletal parts to which the primary and secondary feathers of the wing are connected (Gál 2007, 81, Fig. 13). It is likely that the discarded remains of the black vulture were deposited in a wet layer and covered with organic residue, which resulted in their light brownish color. The 6th–3rd century BC Pantanello Sanctuary included a spring at its heart, along with a constructed collecting basin and channels. The basin silted up in the 5th century BC, but was cleared in the Hellenistic period, when these bones were deposited and again covered (Carter 1990, 416-419, Fig. 6).

The black vulture is a resident and dispersive bird. It became an extinct species in many European countries, including the Apennine Peninsula. Reports of recent sightings mostly come from Spain and southeast Europe (Cramp 1998).

The only domestic bird identified from the Metaponto assemblages is the domestic hen. As Bökönyi mentioned in Chapter 1, the remains found in the 6th century BC feature at Incoronata represent one of the first hen finds in Italy. Early evidence for this species was also found at the 6th century BC site of Paestum near Naples (West and Zhou 1988, 521, Table 2), and the 6th–5th century BC site of Monte Maranfusa in Sicily (Di Rosa 2003, 399-400, Table 2). The latter data reinforce Bökönyi's position and follow Gandert's hypothesis that Greek colonists introduced the domestic hen to Italy. This agrees with Garcia-Petit's opinion about the introduction of domestic hen in the western Mediterranean. The earliest evidence for this domestic bird in Catalonia comes from the 6th century BC, the earliest Greek settlement at Empúries and the nearby Iberian village at L'Illa d'en Reixac (Garcia-Petit 2005, 156). Coeval with the hen bones from the Pantanello Sanctuary are the hen remains from 4th century BC contexts at Roccagloriosa (Bökönyi 1990, 329).

Domestic hen became a frequent and well distributed species all over Europe by the Roman period. Romans apparently introduced conscious breeding and size selection into hen husbandry (Bökönyi 1985, 93). Comparison of measurements from specimens from the chora of Metaponto, dated to the Roman period, with the sizes of coeval finds from Pannonia[2] and Dacia[3] shows that the Italian specimens fall within the average size of hen from eastern regions. One of the two femur measurements and both humerus measurements from San Biagio could represent a male, or they could have come from a larger individual *(Fig. 4.4)*.

As already mentioned in Chapter 1, an injured tarsometatarsus was found in the Pantanello Sanctuary. Judging from the slightly bent bone shaft, this skeletal part most probably suffered a crack or a simple fracture, which was followed by healing without complications *(Fig. 4.2, 1.43)*. Avian remains displaying pathological conditions are rarely found in archaeological bone assemblages. Mechanical traumas, however, are the most frequently encountered lesions both among wild and domestic birds (Gál 2008d). Recently, numerous cases of simple and heavy fractures in domestic hen bones have been reported. Similar conditions of mechanical origin, together with some infectious and metabolic diseases, may result from the improper handling of poultry. Intraspecific conflicts (between birds of the same species), however, may represent another potential source for bone fractures and other injuries. Males of certain species, including galliforms, often fight over their territories and females (Gál 2008e).

[1] *Nat.* 29.25 English translation, John Bostock and H. T. Riley, *The Natural History of Pliny* (London, H. G. Bohn, 1855–57).

[2] Most of these measurements originate from Sándor Bökönyi's unpublished database. I thank László Bartosiewicz for sharing these data.

[3] Gál 2005, 313, Table 5.

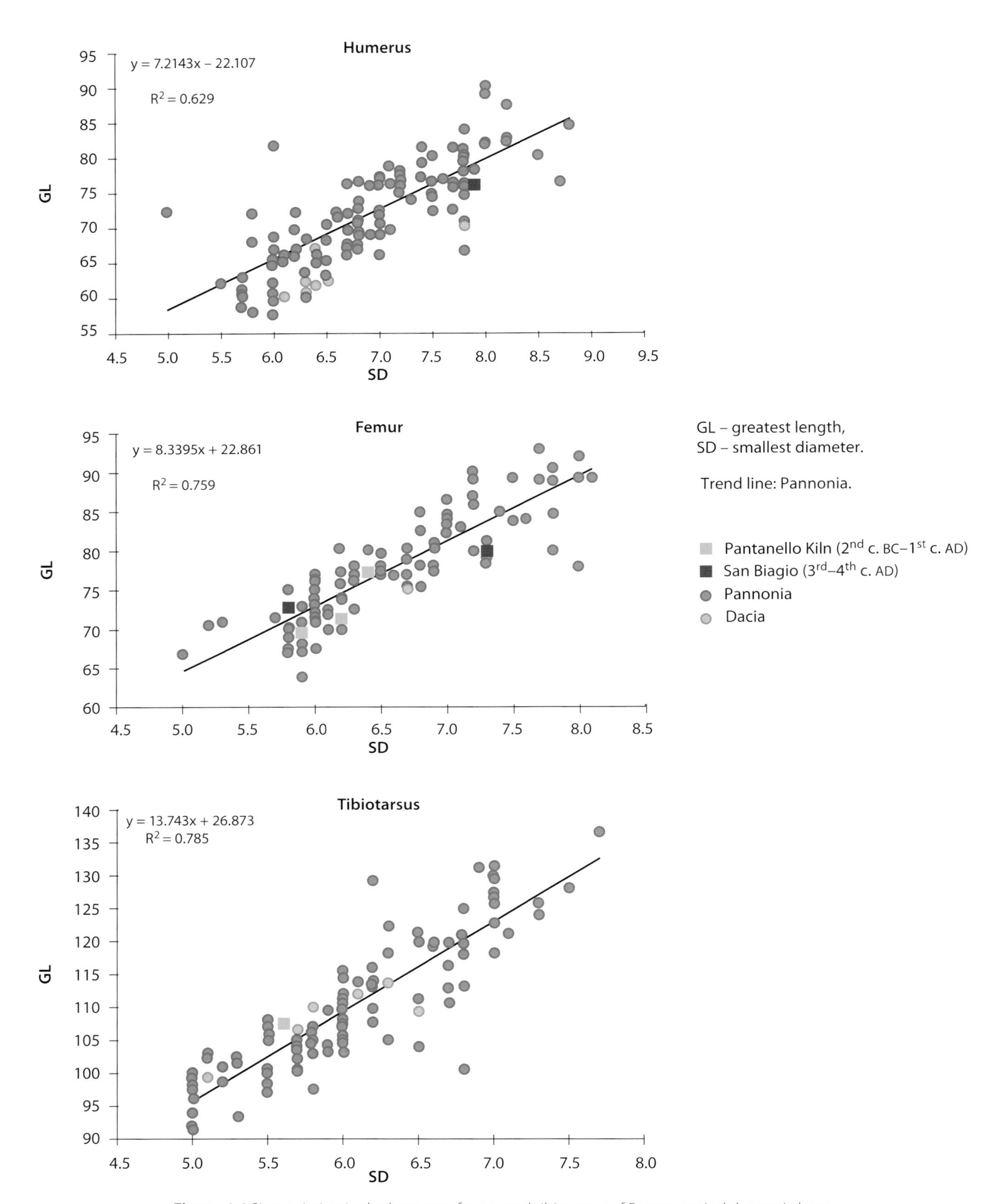

Figure 4.4 Size variation in the humerus, femur, and tibiotarsus of Roman period domestic hens.

Conclusions

Until the Iron Age, the only available birds were wild fowl, which required hunting or trapping. This changed when the Greek colonists introduced the earliest domestic species, the hen. The varieties of wild bird remains—including a water, a steppe, and an arboreal species—indicate a mixed environment around Pantanello during the Neolithic. Various microenvironments, such as pools surrounded by reed-beds, grassland, and forests, could be exploited for a number of wild mammals and birds. Since mallard, partridge, and rook all are resident species in south Italy, their hunting did not have to be a seasonal activity.

It is likely that mallard and partridge were hunted for their meat (and perhaps feathers), but the consumption of rook cannot be excluded either just because this species is not widely considered edible by modern standards. Rook and crow, however, are commensal species that have cohabited with people for a long time. These omnivorous birds commonly occurred in the proximity of human settlements, and may sometimes have even been pursued as pests.

As for the partial black vulture skeleton found at the Pantanello Sanctuary, the religious nature of the site and the specific role of vultures in people's spiritual life (evidenced by several archaeological finds both from Mediterranean Europe and the Near East) suggest that its exploitation was not economic in motive. The missing wing bones indicate that either the wing itself or the separated large feathers may have been used for ritual activities. Evidence for using the skeletal parts of black vultures (or other birds) as raw material is missing both from the avian remains and bone tool assemblages of the Metaponto chora.

After the introduction and spread of hen in animal husbandry by the 6th century BC, it is likely that this species primarily supplied people with bird meat and eggs. Hen remains tend to appear with increasing frequency in Roman period bone collections and already made up 5% of the assemblage from the 3rd–4th century AD site at San Biagio. The period of the Roman Empire was significant in developing breeds and animal trade, and marked the improvement and diffusion of poultry stocks all over Europe, in addition to a number of consciously bred domestic mammals.

The under-representation of avian remains in the animal bone assemblages found in the chora of Metaponto suggests that birds had a marginal economic role during prehistoric and historic times alike. However, as noted at the outset of this chapter, the delicate structure of most avian bones and the method of archaeological retrieval greatly influence the numbers in these assemblages. Other forms of evidence, such as artistic representations *(Fig. 4.5)* and literary references, may help to form a more complete picture of the relationship between the inhabitants and this fragile form of animal life.

Figure 4.5 Detail of a 5th c. BC bronze mirror decorated with doves, from Tomb 354, Pantanello Necropoleis.

5
Bone Artifacts from the Chora of Metaponto

Erika Gál

In addition to the large assemblage of food remains, the sites in the Metaponto chora yielded a number of artifacts and tools made from various animal skeletal parts. Study of these objects may reveal the exploitation of raw material resources, the typology, the manufacturing techniques, and the continuum in each period. Interpretation is equally important, because the various bone tools reflect hidden details of everyday life, from agricultural activity through gambling to sophisticated furniture design. Zoological evaluation as an integral part of multidisciplinary analysis may thus shed light on the technological development and sometimes subtle meaning of these objects.

Vertebrates can be exploited for hard raw materials in addition to soft tissues such as meat, fat, and leather. Bone, antler, and teeth were used in manufacturing in various ways as early as the Late Pleistocene. Among these raw materials, however, antler is the only one that can be acquired from live animals. This opportunity occurs when stags lose their antler racks at the end of the mating season in late winter. Procurement of bone and tusk, by contrast, always involves killing the animal. Therefore, the selection itself of animals already offers information on hunting and animal husbandry, palaeoecological resources, and handicraft. The latter is often connected to the most exploited species at the settlement.

The literature on bone artifacts has become increasingly rich, describing utensils from all over the world. Therefore work must be focused on clearly defined regional and chronological units when interpreting new results.[1] Since a number of bone tools in this study come from prehistoric features, the use of Jörg Schibler's reference works concerning the morphological and typological systematization of Neolithic bone tools from Switzerland (Schibler 1980 and 1981) was considered an appropriate background for evaluating the artifacts from Metaponto. Several papers on food and bone tool assemblages from Italy (e.g., Erickson 1998; Gal 2008f; Giomi 1996; Tagliacozzo 1994 and 2005-2006) were also used for comparing the new data with previously known tool types from this area of the Mediterranean.

The theoretical work done by Alice M. Choyke on the manufacturing continuum also proved important in appraising the character and level of craftsmanship applied to the artifacts (Choyke 1997). Choyke distinguished between Class I and Class II tools by analyzing planned versus opportunistic features of artifacts. Class I tools demonstrate characteristics indicating that they had been planned, starting with the selection of species and skeletal parts available. This consciousness is typical throughout the whole process of manufacturing and patterned use of Class I tools. Such ornaments and utensils reflect the social and economic value of the task they were involved in relative to the society as a whole; therefore, their working life often was prolonged by successive curation. On the other hand, less energy and care were invested in the manufacture of Class II tools, which were often abandoned after completing the task for which they were used. This group also includes ad hoc fragments, which were made from a greater variety of skeletal parts and species, and often used only for a single occasion. Studying individual artifacts within the manufacturing continuum defined between these extremes allows us to interpret cultural attitudes towards bone as a raw material.

What follows is a detailed catalog of the artifacts by chronological order, with classification by Choyke's types as well as the description and state of preservation of each item. The social and economic importance of tasks completed by using bone implements is then discussed by site and period.

[1] László Bartosiewicz and Marta Moreno-García are gratefully acknowledged for their help in collecting relevant references.

Catalog of Bone Artifacts

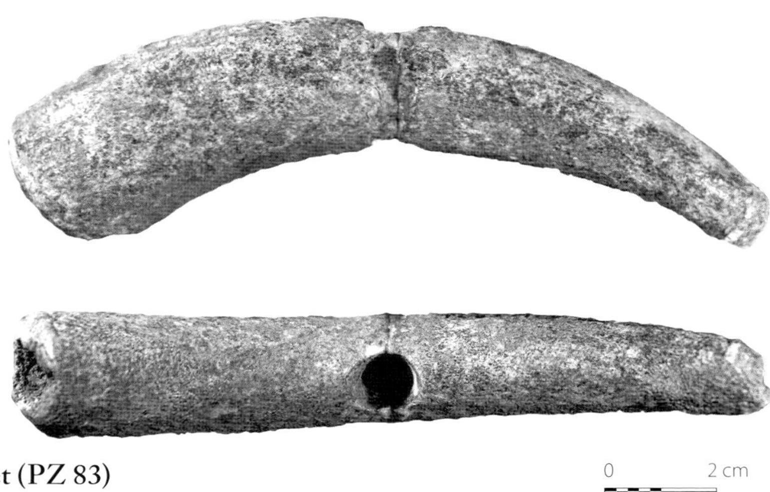

1. Hafted antler socket (PZ 83)
Period: Late Neolithic
Provenience: Pantanello, Pit V
Class: I
Description: The approximately 146.0 mm long (measured along the convex surface) object made from the tine-tip of red deer (*Cervus elaphus* L.) antler has an 8.7 x 6.4 mm hole drilled in its middle part. The cut base of the implement is polished around the edge, which is likely due to fine use wear. There is a 13.2 x 11.3 mm and 6.0 mm deep hole in the spongiosa at this end, which likely encapsulated a stone blade for hammering, while a wood handle fit the drilled hole in the middle (see discussion below, *Fig. 5.1*). The tip of the antler is not worked but was naturally (in vivo) or accidentally broken. The deliberate selection of the raw material as well as the thorough preparation of the implement places it in the group of Class I tools.

---

2. Sheep metapodial point (PZ 83)
Period: Late Neolithic
Provenience: Pantanello, East Pit
Class: I

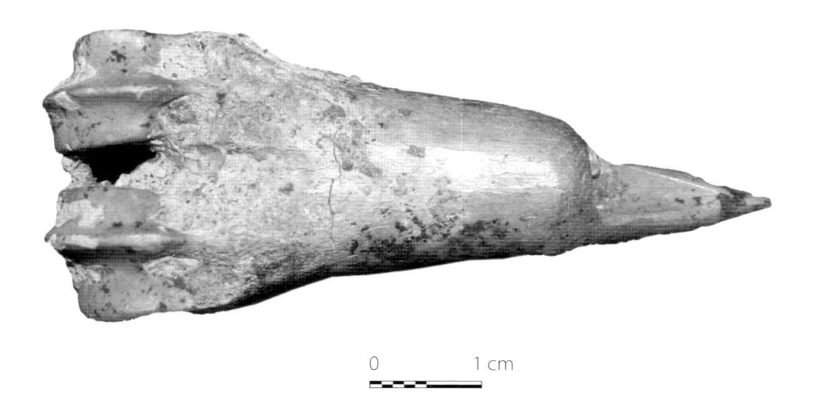

Description: The 62.2 mm long artifact made from a sheep metatarsus corresponds to a small point with articular end (Type 1/4) in Schibler's (1981) classification based on the bone tools from Neolithic Switzerland. The complete distal epiphysis of the metatarsus forms the handle of the point, while the tip was carved from the diaphysis without splitting. According to the heavily polished handle and curated tip, the point had been used over a long time. Similarly to the previous specimen, the selection of raw material and the several planned steps in producing this implement place it among Class I tools. The asymmetric tip of the point is closest to the Form 10/1 in the aforementioned classification (Schibler 1981: 16, Figs. 3-4). The tip of the point is complete. The length of the working tip is 6 mm and its thickness 5.0 mm above the tip is 3.0 mm. The osteological measurements of the metatarsus following the standard by von den Driesch (1976: 92) are: Bd=25.1 mm; Dd=17.7 mm.

3. Bone point (PZ 83)
Period: Late Neolithic
Provenience: Pantanello, Pit I, Level 3
Class: Class I
Description: The 31.7 mm long fragment represents the broken tip of a point. In its present form it corresponds to the Type 1/7 (small point without articular end) in Schibler's (1981) classification, but it may have belonged to the Type 1/1 or Type 1/2 (small ruminant metapodial point). This type of implement is manufactured by several steps of carving that include grooving, splitting, and abrasion. The latter process is highly visible in the rough transversal marks on the whole surface. This is a form of manufacturing wear usually attributed to grinding with sandstone. The greatest breadth and depth of the implement are 9.3 mm and 4.6 mm respectively. The tip corresponds to Form 7/11 (Schibler 1981: 16, Figs. 3-4). The length of the working end is 14.5 mm, while the width 5 mm above the tip is 5.1 mm.

4. Bone point fragment (PZ 83)
Period: Late Neolithic
Provenience: Pantanello, Pit V, Layer 5
Class: II
Description: The 44.4 mm long fragment represents a small point made from a long bone diaphysis of a smaller mammal (possibly caprine). It corresponds to Type 1/7 in Schibler's (1981) classification in its present form. Not much care was taken during the working of the artifact. Only the tip is carved, which places the item among the Class II or ad hoc tools. Its greatest breadth and depth are 7.8 mm and 4.5 mm respectively. The tip corresponds to Form 7/1 (Schibler 1981: 16, Figs. 3-4), and it measures 1.6 mm at 5.0 mm above the end. The working tip of the point is 3.7 mm long.

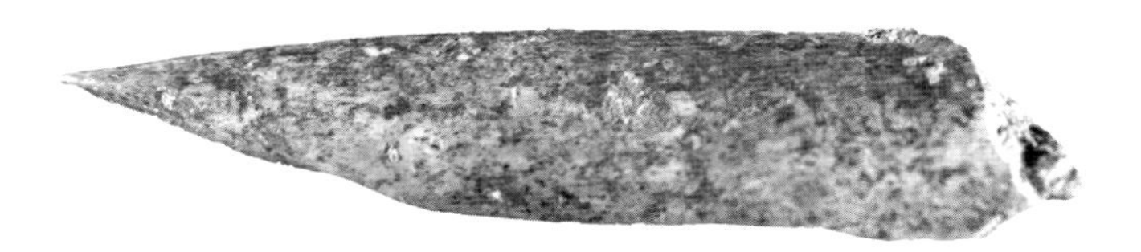

5. Bone point fragment (PZ 83)
Period: Late Neolithic
Provenience: Pantanello, Pit L, Layer 2
Class: II
Description: The 50.1 mm long point represents the tip of a well preserved point made from the tibia of a caprine. The greatest breadth of the implement is 11.6 mm. The length of the tip is 19.3 mm, while its width 5 mm above the end is 2.7 mm. In its present form this Class II artifact corresponds to a middle-size point without articular end, Type 1/8 in Schibler's classification. The top of the implement had been carefully carved (Form 7/1). The lack of damage suggests that this tool was abandoned or lost.

6. Tibia point fragment (PZ 83)
Period: Late Neolithic
Provenience: Pantanello, Pit I
Class: II
Description: This point was made from the diaphysis of a tibia from a small ruminant (roe deer, sheep, or goat). It belonged to a middle-size point (possibly Type 1/8, Class II). The tip was broken and the dense transversal marks on the rather worn surface point to the intensive use of the implement. The length of the point fragment is 83.0 mm, the width is 19.2 mm, and the depth is 13.3 mm.

7. Pin fragment (PZ 83)
Period: Late Neolithic
Provenience: Pantanello, Pit M, Layer 4
Class: I
Description: The 24.4 mm long, 3.3 mm wide, and 2.2 mm thick artifact represents the middle section of a gracile pin, carefully carved out of the compact tissue of a long bone diaphysis of an unknown animal. According to its shape and the transversal marks covering the entire surface, as well as the frequency and form-variety of small ruminant metapodial perforators during the Neolithic (Schibler 1981; Sidéra 2005; Gál, 2008f), it may have belonged to Schibler Type 1/1. It represents a Class I tool.

8. Ruminant metapodial tool fragment (PZ 83)
Period: Late Neolithic
Provenience: Pantanello, Pit A
Class: I
Description: This tool was carved from the diaphysis of a small ruminant metacarpus. According to the straight and slender shape of the object it may have originated from roe deer (*Capreolus capreolus* L.). The long splinter of bone compacta is covered by high polish, although the tip is missing. The two fragments do not match because a matrix containing small stones is cemented to the broken surfaces. The complex character of manufacturing places it among Class I tools.

9. Chisel/spatula fragment (PZ 83)
Period: Late Neolithic
Provenience: Pantanello, Pit A
Class: II
Description: The 51.9 mm long and 8.5 mm wide implement represents the tip of a fine chisel-spatula (Schibler Type 15, Choyke Class II). The peak of the artifact had been carved in a spatula-shape 11.0 mm from the end. The tip had been rounded and sharpened. It corresponds to Form 31/23 within the Neolithic classification (Schibler 1981: 18, Fig. 6-7). The greatest breadth of the chisel tip is 5.2 mm.

10. Bone fragment with fine bevel (PZ 83)
Period: Late Neolithic
Provenience: Pantanello, Pit V, Layer 5
Class: II
Description: The 39.3 mm long bone fragment comes from a thin-walled long bone diaphysis. It represents a used bone fragment rather than a tool. This piece was neither heavily worked nor used for a long time. The length of the working edge is 17.2 mm, while the width is 3.2 mm.

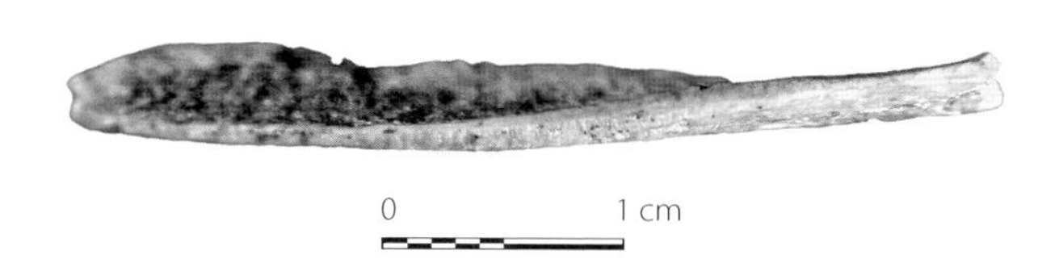

11. Chisel fragment (PZ 83)
Period: Late Neolithic
Provenience: Pantanello, Pit I, Layer 6
Class: II
Description: The 45.3 mm long, 32.1 mm wide, and 11.3 mm thick tool fragment is possibly the broken end of a massive ad hoc chisel (Type 4/7). It was made from the long bone diaphysis of a large animal. Little work was invested in the manufacturing of this implement. The chisel has an irregular shape close to Form 29/24 with a short (16.6 mm wide) polished sequence.

12. Polished bone fragment (PZ 83)
Period: Late Neolithic
Provenience: Pantanello, Pit E, Layer 5
Class: II
Description: The 40.2 mm long specimen showing traces of burning is a diaphysis fragment from caprine ulna. It is not worked but polished, which places it in Schibler's Type 19 classification (1981).

13. Metatarsus point fragment (A80/2/III/XXVII, L, a 365)
Period: Bronze Age
Provenience: Termitito
Class: I
Description: This fragment of a middle-size point without articular end (Schibler Type 1/8) was made from the diaphysis of a red deer metatarsus. Both base and tip are missing. The choice of high quality raw material and marks of manufacturing place it near Class I tools in the manufacturing continuum. The length of the fragment is 54.4 mm, the width is 13.1 mm, and the depth is 10.0 mm.

14. Antler debris (A80/2/III/XXVI/L, a 350; A 80/2/IV/XXIV/L, a 310)
Period: Bronze Age
Provenience: Termitito
Class: debris
Description: Two pieces of antler manufacturing debris were found at Termitito. The larger fragment (14a) was connected to the frontal part of the cranium, evidence that the animal was killed. The smaller piece (14b) is 30.5 mm long and has marks of metal sawing on two surfaces. It may have come from a similar specimen or from shed antler.

15. Metapodial tool fragment (IC 78-191B)
Period: 8th century BC
Provenience: Incoronata, Feature C6, 2-3 m from the NW end of the trench
Class: I
Description: The 72.5 mm long artifact represents the broken handle of a large tool. It was either a massive point (Type 1/6) or a massive chisel (Type 4/3). It was carved out of the distal end of a longitudinally split red deer metatarsus. The axial side of the split bone shows handling polish. The use wear places it among Class I tools in the manufacturing continuum.

16. Metapodial point fragment (IC 78-106B)
Period: 8th century BC
Provenience: Incoronata, proximal part from Pit D, tip from Square A2, level 3
Class: I
Description: This burnt tool was carved out of the diaphysis of a large ungulate metapodium. The two fragments match each other and measure 69.2 mm together. The tip was sharpened by flint and is covered by use wear in the form of high polish. It corresponds to Type 1/6 or 1/9 in Schibler's and Class I in Choyke's classification. The working tip is 29.5 mm long and 4.2 mm wide at 5 mm from its end.

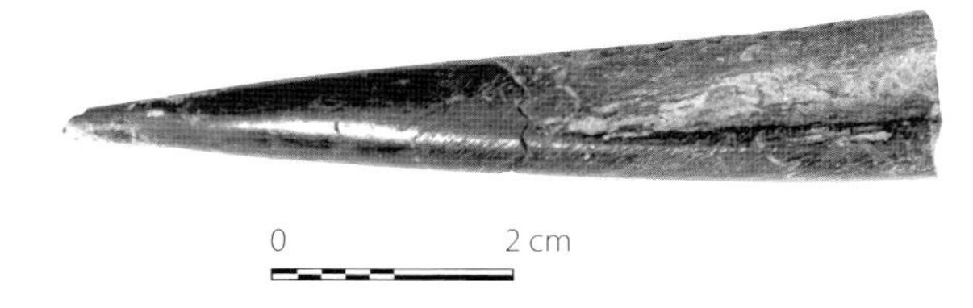

17. Polished astragalus (IC 77-6)
Period: 6th century BC
Provenience: Incoronata, Feature S1, B2
Class: I
Description: Both the lateral and medial surfaces of this 27.3 mm long, left-side sheep astragalus are worn flat. No other modification can be observed on the specimen. According to a number of archaeological and ethnographic parallels, it was probably used as a gaming object.

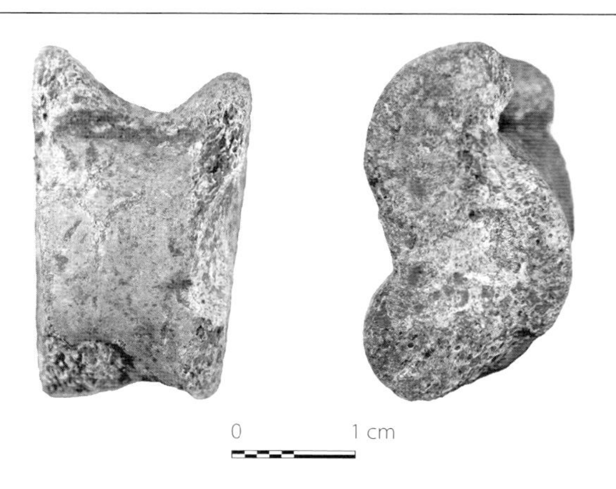

18. Drilled astragalus (PZ 82-310S)
Period: 6th–3rd century BC.
Provenience: Pantanello Sanctuary, carbonated layer in Obj. 5, N'-6, baulk 7;
Class: I
Description: This left-side sheep astragalus was drilled through the articular surface. The edges of the 3.4 mm diameter hole show little wear. The highly polished condyles and wear-marks inside the hole suggest that the astragalus was possibly suspended and in contact with a soft tissue (cloth or skin). The bone could be measured according to the internationally used osteometric standard (von den Driesch 1976: 88) since the other parts were intact: GL=30.6 mm; Dl=16.8 mm; Dm=17.9 mm; Bd=19.4 mm.

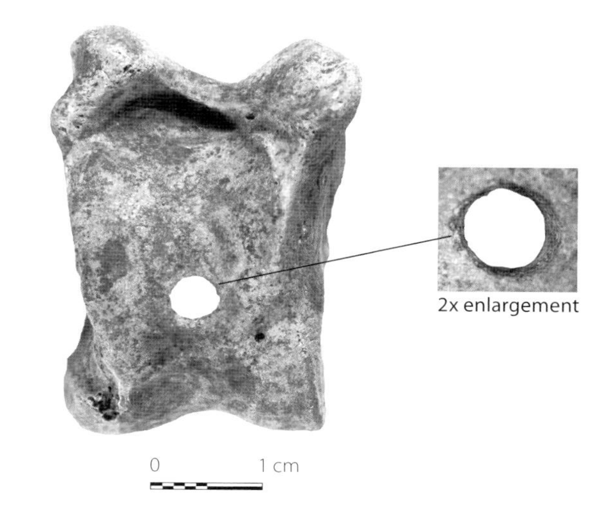

19. Bone handle (PZ 76 344)
Period: 6th–3rd century BC.
Provenience: Pantanello Sanctuary
Class: I
Description: This find likely represents a handle fragment. Both ends were rounded in a cylindrical shape. According to the external and internal structure of the bone, it is likely that a radius or a metapodium of a large mammal served as the raw material for this piece. The greatest length of the fragment is 70.5 mm, the greatest width is 19.4 mm, its depth is 7.7 mm. Two fine ribs were carved in a transverse direction, likely as decoration. They were placed at 11.mm from one end, while the remains of a similar decoration can be detected at 6.5 mm from the other end as well. The thickness of the bone wall is 4.6 mm at one end and 7.4 mm at the other.

20. Antler handle (?) fragment (PZ 77-1134P)
Period: 6th–3rd century BC.
Provenience: Pantanello Sanctuary, Square O'–P' 2-3, level 3
Class: I
Description: The 36.6 x 28.5 x 17.6 mm implement was carved from an antler tine. The tip is missing. The surface is carved and polished. The spongiosa shows holes in both ends of the antler, but it is difficult to distinguish what is natural breakage or artificially made. The wider end of the piece is cut. The base was flattened to triangular surfaces on both sides, with fine polish on the triangular edges on both surfaces.

21. Chisel fragment (PZ 77-1042P)
Period: 6th–5th century BC.
Provenience: Pantanello Sanctuary, Square N'-2, N'-3
Class: II
Description: The 51.0 mm long and 34.2 mm wide fragment represents the broken end of a massive bevel ended tool (Type 4/3, Class II), made from a long bone diaphysis of a large mammal. The edges are rounded (Form 29/23). The length of the working edge is 28.5 mm, while the width is 31.1 mm. Soil matrix is cemented in the marrow cavity. The compact surface holds breakages from heavy use suggesting this piece may have been used for chiselling.

22. Antler debris (PZ 76-391; PZ 78-549)
Period: mixed Greek-Roman; 6th–3rd century BC
Provenience: Pantanello Sanctuary,
Square L'-1, level 1 (22a);
Square M' 6-7-8, level 3 (22b)
Class: debris
Description: Both pieces represent debris from an antler workshop, but differ in their mode of procurement. The first specimen (22a) is a skull fragment with antler, which means that the animal was killed. According to the diameters of the antler rose (72.5 mm x 65.0 mm), this fragment belonged to a well developed stag that was likely killed not only for antler but also meat and other materials. Both the eye-tine and the main beam of the antler were cut off. The second piece (22b) is the base of a shed antler. Similarly to the first specimen, both the eye-tine and the main beam were cut off. The diameters of the antler rose (74.1 mm x 68.7 mm) suggest an even larger stag. See also Figure 1.8 in Chapter 1.

22a

22b

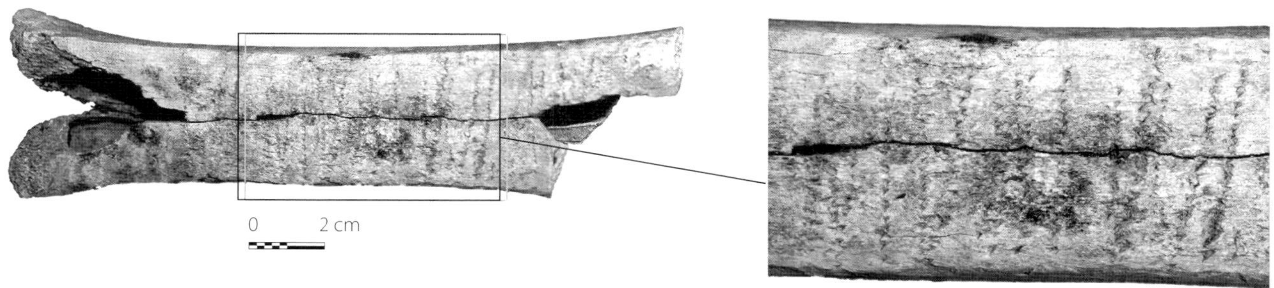

2x enlargement showing serrated sickle marks.

23. Bone anvil (PZ 80-191B-1)
Period: 2nd century BC–1st century AD
Provenience: Pantanello Kiln Deposit, Square J'-1 E
Class: I
Description: This object was made from a cattle metacarpus. The distal part of the artifact is missing. Both the dorsal and palmar surfaces and the proximal epiphysis of the bone were cut flat. The lateral and medial surfaces were also modified. Less modification was necessary on the palmar surface, while the major portion of the dorsal surface was removed to produce an oblong section. Therefore it is mostly the palmar surface that bears special marks, caused by the maintenance of serrated sickles. The greatest length of the implement is 161.5 mm. Its greatest breadth at the distal end is 49 mm, while its greatest depth at the same end is 32.2 mm.

---

24. Semi-cylindrical bone furniture fitting (PZ 81-699B)
Period: 2nd century BC–1st century AD
Provenience: Pantanello Kiln Deposit, NW trench, Batt 7, brown sand/gravel, carbonated layer.
Class: I
Description: The thoughtfully worked artifact was made from the metatarsus diaphysis of a large ungulate, most likely cattle. The 89.4 mm long semi-cylinder displays 11 longitudinal parallel, cannelure-type grooves. Its wider end measures 27.8 x 15.2 mm, while its narrower end is 25.0 x 13.1 mm. All edges were carved and polished. The artifact shows three brownish patches which are probably traces of burning. Two fragments of similar fittings were found at the site of Sant'Angelo Grieco (Catalog No. 29).

---

25. Flat bone inlay (PZ 81-658B-11)
Period: 2nd century BC–1st century AD
Provenience: Pantanello Kiln Deposit, NW trench, E end, red-brown soil
Class: I
Description: The rectangular plaque (48.1 x 10.4 x 2.7 mm) was carved from the diaphysis of a large ungulate long bone or even rib. One of the ends is broken, showing that the original length of the object was greater. All five undamaged surfaces were polished.

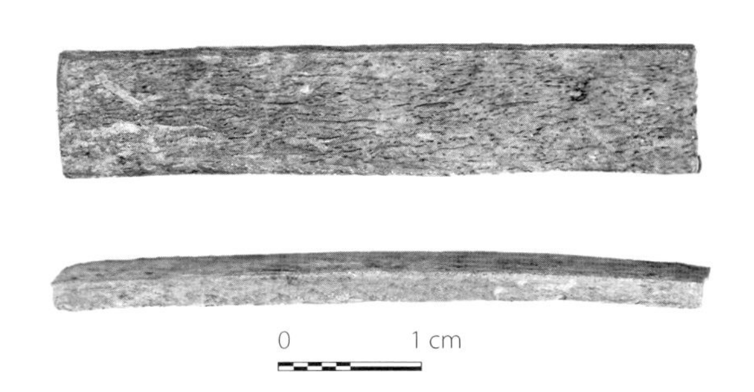

26. Short hinge (PZ 75-239x; PZ 75-652x)
Period: 2nd century BC–1st century AD
Provenience: Pantanello Kiln Deposit, Square B' 11 W; Sounding W, Omega trench south, square C' 8
Class: I
Description: The hinge was carved out from a bovine (*Bos taurus* L.) metatarsal diaphysis. The greatest length is 28.1 mm, the greatest breadth is 25.9 mm, while the greatest depth is 12.0 mm. A 7.4 x 7.1 mm hole was drilled perpendicularly, roughly in the middle of the bone wall. The artifact had been broken, probably due to usage, and was found in two pieces in different coeval features. See Fig. 5.6.

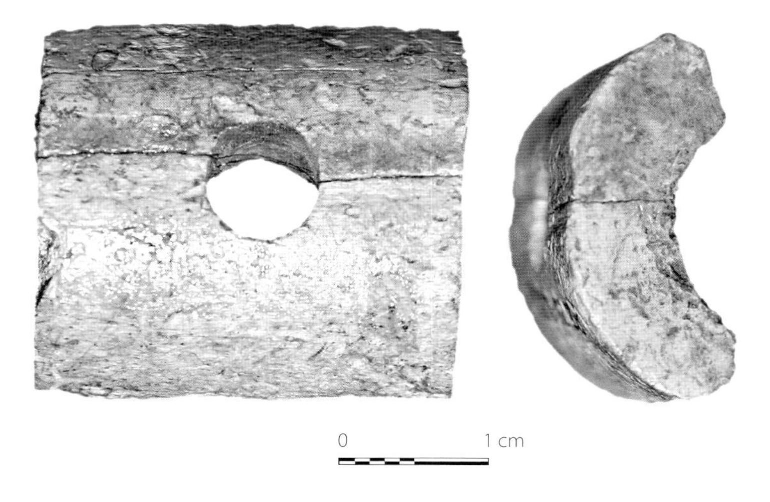

27. Bone pivot (PZ 81-81B)
Period: 2nd century BC–1st century AD
Provenience: Pantanello Kiln Deposit, upper level of ramp to trench
Class: I
Description: According to its shape, the bone ring was carved from a metacarpus diaphysis of a small equid. The object (30.0 mm wide, 25.4 mm deep, and 8.1 mm thick) was made by sawing and heavy polishing of the section. The wall of bone measures 6–7 mm. Both flat surfaces of the ring show marks of circular polishing, due most probably to the frequent moving of the corner joint of a small furniture piece or chest.

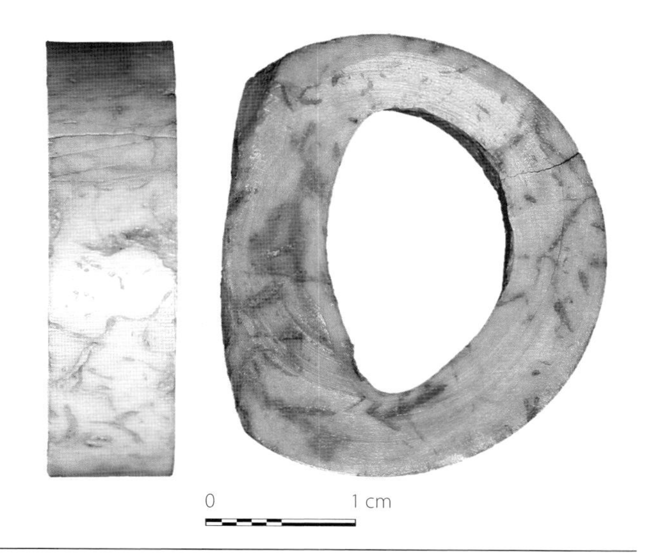

28. Sawed antler handle with iron insert (PZ 81)
Period: 2nd century BC–1st century AD
Provenience: Pantanello Kiln Deposit, NW trench, Batt. 6
Class: I
Description: This object has two parts. The first, 29.3 mm long, is a sawed and drilled piece of antler beam forming the handle. The greatest diameters of the antler cylinder measure 27.7 mm x 22.9 mm. The second part, an iron insert of approximately 10.5 mm diameter, was set into the drilled spongiosa of the carved antler. This iron object fills the antler fragment in its entire length.

29. Fragments of semi-cylindrical bone furniture fittings (SGI 81-58B; SGI 81-98B)
Period: 2nd century BC–1st century AD
Provenience: Sant'Angelo Grieco, Sounding I, Batt. 3, area II
Class: I
Description: Both fragments resemble the object found intact in the Roman period Kiln Deposit of Pantanello (see Catalog No. 24). They were not only broken, but show more erosion than the specimen from Pantanello.

30. Antler ring (SA 79-1007 B)
Period: 2nd century BC–1st century AD
Provenience: Sant'Angelo Grieco
Class: I
Description: The heavily eroded, round antler ring was carved from a thin and rounded section of an antler tine. The outer surface of the compact tissue was almost completely removed. The outer diameter of the ring is 20.8 mm, the inner diameter measures between 11-12 mm. A hole, 4.9 mm in diameter, was drilled perpendicularly to the long axis, and may have served for mounting a piece of metal or glass inlay.

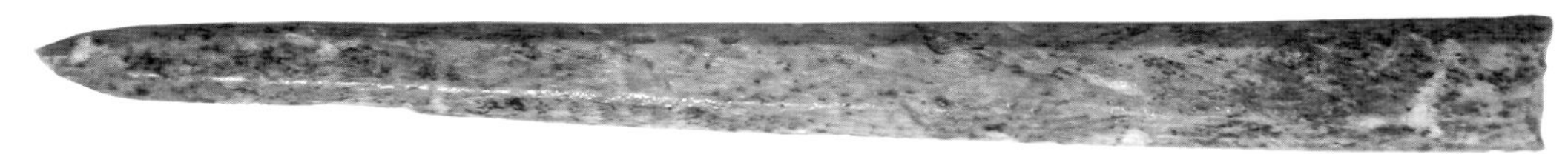

0 0.5 cm

31. Perforator fragments (SB 80-372B; SB 80-247B)
Period: 3rd–4th century AD
Provenience: San Biagio, Hypocaust, Batt. 2 (N half); Hypocaust, level 3, Batt. 1.
Class: I
Description: Both longitudinally faceted objects were carved from long bone diaphyses. The longer fragment, 44.9 mm in length, represents the tip of a fine perforator. The working tip is 6.3 mm long and 2 mm wide and worn. Both ends of the shorter fragment, 36.0 mm long by 3.3 mm wide, were recently broken. The second fragment is not shown due to its poor state of preservation.

Discussion

The bone artifacts from the chora of Metaponto represent great variety in both chronological and typological terms. The most abundant tool collection comes from the Neolithic pits at Pantanello, where twelve objects, made of bone and antler, were found in the earliest settlement within the site complex. These represent 0.9 % of the whole Neolithic bone assemblage *(Table 5.1)*. The abundance of bone artifacts from this period is not surprising, since the skeletal parts of animals must have been a highly appreciated raw material before the invention of metallurgy.

From a typological point of view, points represented more than half of the tool assemblage. They were used primarily for perforating soft substances such as leather and textiles. The majority of points were carved from sheep and/or goat metapodia. Except for the best preserved point where the complete distal epiphysis that formed the handle survived (Catalog No. 2), only tips or middle sections were found. Judging from their narrow shape, these were carved from longitudinally split metapodium fragments. The manufacturing of these small ruminant metapodial points was widespread during the Neolithic all over Europe (e.g. Camps-Fabre and D'Anna 1977; Murray 1979; Beldiman 2002; Sidéra 2005). The majority of artifacts from other Neolithic sites in central and south Italy also belong to this type (Giomi 1996; Tagliacozzo 2005-2006; Gál, 2008f).

Two-thirds of the twelve zoologically identified points came from sheep and goats. This proportion corresponds to the taxonomic distribution of food remains at the Neolithic pits at Pantanello, where the great majority—81%—of the identified skeletal parts belonged to caprines (Chapter 1). Red deer, a game animal commonly used for many Neolithic materials, was surprisingly under represented in both food and tool assemblages. Only one antler fragment was found in each collection. Since shed antler may be collected without killing the animal, one may conclude that red deer was not abundant or people did not hunt it at Pantanello during the Late Neolithic. The few remains attributed to a great number of wild species at this site indicate that hunting generally did not play an important role in the economic life of the settlement, and must have been of an opportunistic nature.

Nevertheless, the antler-socket found in Pit V at Pantanello (Catalog No. 1) is a special find. It represents one of three hard pieces needed for making a hafted implement *(Fig. 5.1)*. Its manufacture involved the procurement of an antler, cutting the tine to size, and drilling. The carefully drilled hole through the middle section of the specimen from Pantanello indicates that a handle—most probably wooden—was fitted into the antler. Also, the hollowed cavity at the thicker end of the socket suggests that another piece—most probably a stone blade—was wedged into it. The highly polished surface around the hole suggests that

Site Name	Period	Total Animal Bones	Class I Tools	Class II/ ad hoc Tools	% tools
Pantanello	Late Neolithic	1,360	5	7	0.88
Termitito	Late Bronze Age	692	1	2	0.43
Incoronata	8th–6th century BC	1,668	3	0	0.18
Pantanello Sanctuary	6th–3rd century BC	1,186	3	3	0.50
Pantanello Kiln Deposit	2nd c. BC–1st c. AD	1,598	6	0	0.37
Sant'Angelo Grieco	2nd c. BC–1st c. AD	125	3	0	2.40
San Biagio	Late 3rd–4th c. AD	570	2	0	0.35

Table 5.1. Number and distribution of bone artifacts by site and class, Metaponto chora.

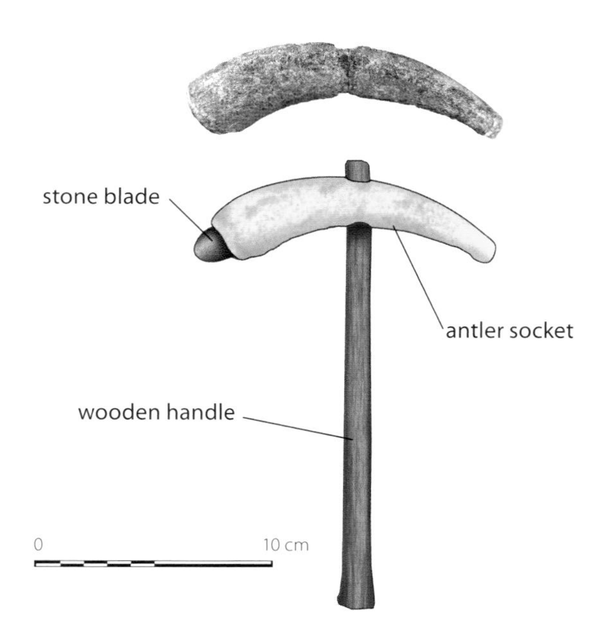

Figure 5.1 Reconstruction of hafted implement (Catalog No. 1), Neolithic Pantanello (LB/DB).

this end of the object must have been often in contact with a soft material. The closest parallel to this object comes from the Cortaillod Culture (circa 3900–3500 BC) of the Swiss Neolithic (Suter 1981, Table 71, Figure 1061). Similar hafted implements made from antler and identified as hatchets were described at the Early Chalcolithic site of Drama in Bulgaria (Sidéra 1996: Plates 5-6), among others. However, the relatively small hafting hole in the specimen under discussion here is indicative of a relatively fine, hammer-like instrument with a thin and probably not very long handle. The Early Neolithic site of La Marmotta in Central Italy also yielded handles made from red deer antler (Tagliacozzo 2005–2006: 433).

Antler was a commonly used raw material in prehistoric times. Softer but more flexible than bone, it is well suited for producing complex tools. A great variety of implements made from antler have been described from Neolithic sites in Central Europe (Winiger 1999; Zalai-Gaál and Gál 2005). A number of cranial remains and tine ends from red deer found at the Mesolithic-Neolithic site of Grotta dell'Uzzo in Sicily indicate that antler had been removed from the skull but most probably worked outside the cave, since no evidence of antler working was found at the site. Cut marks above the antler rose and the "ready" tine points suggest that people cut off the antlers at the hunt site, carried the body together with the skull to the cave, and then brought in the cut-off tine ends. The declining number of red deer remains during the Neolithic phases of the site is noteworthy in regard to the Mesolithic and transitional periods (Tagliacozzo 1994). The decreasing trend could have indicated the saving of energy in transport, since the heavy bodies of these animals had to be carried from a greater distance. On the other hand, the decline in red deer hunting (and that of other wild animals) fits in with the increased share of sheep and goat remains found in the food refuse, and generally with the spread of animal husbandry during the Neolithic.

The assemblage from the Bronze Age site at Termitito yielded only two worked objects. In addition to the poorer representation of skeletal parts in this assemblage, the decrease in the use of hard raw materials of animal origin must have been connected with the invention of bronze and the beginning of metallurgy. Interestingly, both the metapodial point and antler-debris come from red deer, which also yielded 90% of the skeletal parts from wild animals in this small refuse bone sample.

By the beginning of the Archaic period (8th–6th century BC), the number of fine artifacts increased notably in the chora of Metaponto *(Table 5.1).* Two objects from 7th century BC layers at Incoronata are still metapodial tools (Catalog Nos. 15–16), but they were manufactured from the bones of large ruminants instead of sheep and goat. Cattle was the third best represented domestic animal in the small food assemblage from this site. Red deer predominated by 78% among wild animals, but made up less than 5% of the whole bone assemblage (Chapter 1).

The third carved object from Incoronata was a polished sheep astragalus (No. 17). It was found in a 6th century BC context and may represent a strong Greek influence or even an imported object. The use of astragali in cultic life has a long tradition. One of the earliest instances comes from two Copper Age sites in Romania. The Cucuteni B site of Ghelaieşti-Nedeia yielded 497 astragali from caprines and pig, which were found in a pot in a sanctuary. Some of these objects were polished and others were incised. Two ritual deposits were found in the Cucuteni A tell site of Poduri-Dealul Ghindaru, where two astragali were found together with a cup of seeds and other possibly cultic objects in one of the contexts. Traces of abrasion, fire, and brown-reddish ochre could be identified on the astragali from cattle and caprines in the other feature (Bejenaru and Monah 2004). Other examples of the ritual role of astragali include finds from the Copper and Bronze Age in Anatolia. An astragalus made of gold was found in a grave in Varna (Kovács 1989: 103, citing Ivanov, I. S.).

The use of astragali as gaming pieces likely dates back to the Bronze Age, as evidenced by the Early and Middle Bronze Age sites of Malatya-Arslantepe in Anatolia and Jászdózsa-Kápolnahalom in Hungary (Erickson 1998: 838–839; Bartosiewicz 2006: 192). The latter site yielded a ground down red deer astragalus (Csányi and Tárnoki 1992: 196, Fig. 291/5), similar to the polished specimen from Incoronata. Spread of this entertainment may be due to Greek influence, as shown by a number of references and representations (Bartosiewicz 2006: 192). The early pieces simply took advantage of the natural cube-like shape of astragalus; some were ground down in order to emphasise this form, as in the case of the Pantanello specimen. Some later finds display various grooves, dots, and holes on the smooth posterior articular surface, which actually forms the joint between the astragalus and the heelbone or calcaneus (e.g., the Roman period Sarmatian cattle astragalus from the site of Gyoma 133 in the Great Hungarian Plain (Choyke 1996: 321, Fig. 5) or the late medieval sheep-astragalus collection from Szolnok in central Hungary (Kovács 1989).

Holes drilled into gaming pieces made from sheep astragali were sometimes filled with lead in order to provide weight to the dice. A drilled hole may be also noted on the posterior articular surface of the sheep astragalus found at the Pantanello Sanctuary. Traces of lead or other material could not be noted in or around the hole, the margin of which was slightly damaged (Catalog No. 18). This, along with the polished opposite surface of the bone, would suggest that the drilled astragalus was suspended and worn as an amulet. Ruminant astragali evidently had broad cognitive and symbolic meanings. A large number of specimens were found in five burials—one containing 50—at the Pantanello Necropoleis, mostly in tombs belonging to children or infants *(Fig. 5.2)*. These examples are discussed in the larger context of grave goods by Erickson (1998: 838–839; also Bökönyi 560–561). A large bronze copy of an astragalus, with two handles and bearing five lines of a dedication in Greek, was found at the ancient site of Suse in modern-day Iran. According to the type of letters, it was possible to date the artifact to the third quarter of the 6th century BC (André-Salvini and Descamps-Lequime 2003). This suggests Greek influence as a possible link between at least some of the astragalus finds.

A chisel fragment, a bone chisel, an antler handle, and antler debris were also found in the Pantanello Sanctuary. The cavity drilled in the proximal part

Figure 5.2 Astragali in a 5th century BC grave of an infant. Pantanello Necropoleis (Tomb 264).

Figure 5.3 Bone amulet with a carved bust of a warrior, from a mid-5th c. BC female burial. A hole on the backside allowed the amulet to be worn. Pantanello Necropoleis (Tomb 350).

of the worked antler piece (Catalog No. 20) and the polished surface on both sides would suggest that the object was the handle of some fine tool. The presence of an antler workshop is also attested by the piece of antler-debris that lacks both the eye-tine and main beam (Catalog No. 22). A similar antler piece was found in the Kiln Deposit at Pantanello (Catalog No. 28), but this object also contained an iron insert.

Two particularly fine objects—a bone amulet (Fig. 5.3) and a lyre made from a tortoise carapace–(Fig. 5.4–5)—were found in 6th and 5th century BC graves at the Pantanello Necropoleis. These too are discussed in detail by Erickson (1998: 838) and Prohászka (1998: 820-821).

The carved bone anvil (Catalog No. 23) found in the Pantanello Kiln Deposit was connected to agricultural harvesting. The rough horizontal lines displayed on the flattened cattle metacarpal bone are marks left by saw-teeth from serrated sickles. Ethnographic parallels from the Iberian Peninsula indicated that tooth deformations of metal sickles were smoothed out by holding the anvil between the knees and hitting the bent section of the sickle against it. A great number of horse and cattle metapodium anvils (as well as cattle mandible anvils) from 10th–20th century sites in Morocco, Spain and Portugal offer parallels for this unusual find (Nadal and Roure 2004; Moreno-García et al. 2005). According to the latest publication on this subject (Moreno-García et al. 2007), dromedary bones also were used as such. The earliest evidence for bone anvils, described as bone rasps or files, comes from the Hellenistic (4th century BC–1st century AD) site of Olbia in south-central Ukraine (Semenov 1964). The same author referred to the existence of similar finds from Neapolis and Phanagoria, the Graeco-Scythian area around the Black Sea. Consequently, the Pantanello find dated to the 2nd century BC–1st century AD is probably the oldest bone anvil of this type in Europe. Within the chronological and regional distribution of metapodial anvils, the Metaponto specimen fills a gap between the eastern Hellenistic regions and the Visigothic Period (5th–6th century AD) in Spain.

Other special objects found in the Kiln Deposit are bone furniture fittings and parts of hinges. One long, semi-cylindrical and highly worked object (Catalog No. 24) formed the cover for a furniture leg. The closest known parallel for this artifact comes from Augusta Raurica, Switzerland. Some of the twenty-seven long covers (Deschler-Erb 1998: 403, Table 51) display the same fine longitudinally carved ridges as may be seen on the object found in the Pantanello Kiln Deposit. The rectangular bone plaque (Catalog No. 25) was most likely a piece of inlay for a flat furniture surface (e.g., desk, bed, or chest).

Figure 5.4 Carapace from a land tortoise used for lyre, found in a 6th century BC grave. Pantanello Necropollis (Tomb 336).

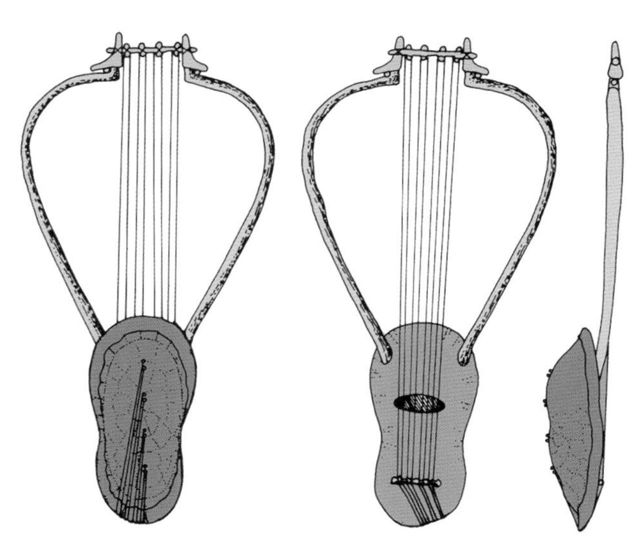

Figure 5.5 Reconstruction of the tortoise shell lyre. Bronze tacks and iron fittings were used to hold the stretched ox hide that covered the sound box and bracing for the arms. Pantanello Necropollis (Tomb 336). (A. Toxey/ICA)

Another small drilled bone plaque (Catalog No. 26) was likely a short hinge used as a fitting for a door or lid on a piece of furniture *(Fig. 5.6).* Bone hinges were common during the Roman era. A number of various short and long hinges were presented by Sabine Deschler-Erb in her book on Roman artifacts from Augusta Raurica, where short pieces predomi-

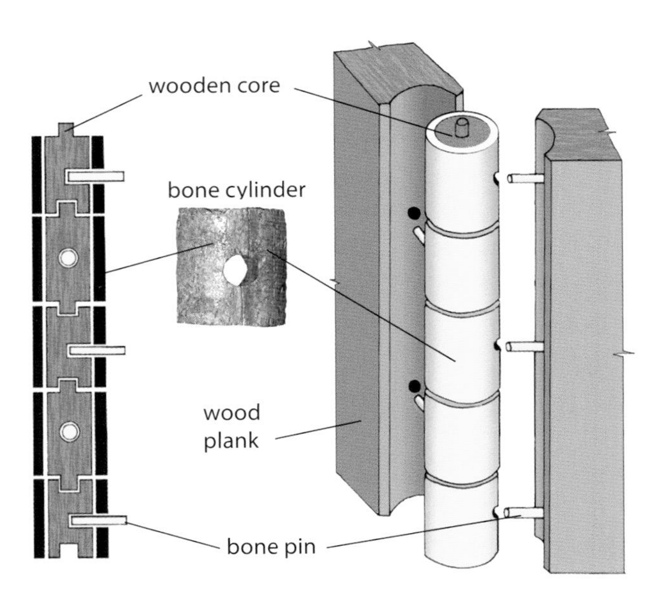

Figure 5.6 Cross section and reconstruction of a bone hinge. Rotation took place between the alternating bone cylinders fixed to the planks (redrawn by LB/DB after Béal 1984).

nated in the hinge assemblage, which included 376 items. Short hinges below 30 mm, however, like the one from Pantanello (28.1 mm), were very rarely employed at Augusta Raurica (Deschler-Erb 1998: 182–189, 398-399, Tables 46–47).

The form and high polish of the bone ring (Catalog No. 27) also indicates a special item. Both flat surfaces of the object displayed concentric and polished abrasion. Since these kinds of marks could be left by the cylindrical part (axis) of an opening structure, it is likely—and its shape also would confirm—that this article was part of a corner joint.

The two under-represented bone assemblages from the Roman period sites of Sant'Angelo Grieco and San Biagio yielded only a few bone artifacts. Nevertheless, the two carved semi-cylindrical fragments (Catalog No. 29) from Sant'Angelo Grieco represent important finds, since they resemble the completely preserved furniture leg cover from the Pantanello Kiln Deposit. The antler ring is also an unusual find. A metal or glass inlay fixed in the small hole may have decorated this ornament (Catalog No. 30). The two longitudinally faceted fragments from San Biagio (No. 31) were probably used as fine perforators or pins, and prove the use of animal bone tools in the chora of Metaponto as late as the 4th century AD.

Conclusions

The bone tool assemblage from the various Metaponto chora sites provides a unique opportunity to review changes in the use of animal-derived implements and artifacts across a time span of approximately 4,500–5,000 years in the region. The taxonomic representation of animals in the bone tool assemblages and food remains tends to be congruent at prehistoric sites. Skeletal parts of sheep and goat predominated in the Neolithic pits at Pantanello, while red deer was especially exploited at Bronze Age Termitito.

The arrival of the Greeks prompted a transition from uniform everyday tools towards more sophisticated and specialized objects created by craftsmen in workshops. This trend continued with the Roman occupation in the region. The identified bone tools either represented fine and decorated objects—such as bone or antler handles—or were employed in special activities like the bone anvil from the Roman period Kiln Deposit at Pantanello. It is likely that the latter object is one of the earliest specimens introduced to Europe from eastern Hellenistic regions. It also offers indirect evidence for the use and curation of serrated sickles, an agricultural innovation that increased the efficiency of harvesting cereals.

Other artifacts, such as the astragali used as gaming pieces or amulets, represent a widely spread but only partially understood relation between gambling and more mysterious meanings. The bone inlays, covers, and hinges often used in decorating furniture indicate a more sophisticated use of bone as a raw material in composite structures. Judging from the various special implements and artifacts which followed the use of simple prehistoric bone points and chisels, the careful selection and manufacturing of bones and antler must have been part of the more refined technology and aesthetic influence of the Greek and Roman people who later populated the Metaponto area.

Appendix: Bone Measurements

The Appendix contains data on all bones which could be measured. Measurements were not available for the bones from the Metaponto Sanctuary (cloaca) nor those from the Greek Pit at Pantanello. All abbreviations are after von den Driesch (1976) unless spelled out, and measurements are in millimeters.

Species	Skeletal part	GL	Greatest d		Smallest d		Base circum.		Site name	Period
Bos taurus	horn core	340.0	80.0		64.0				Pantanello Kiln Deposit	2nd c. BC–1st c. AD
Bos taurus	horn core		46.0		36.0		140.0		Incoronata	8th–6th c. BC
Bos taurus	horn core		57.0						Pantanello Sanctuary	6th–3rd c. BC
	Skeletal part	**P2–P4 length**	**M1–M3 length**		**M3 length**				**Site name**	**Period**
Bos taurus	maxilla		77.0						Pantanello Sanctuary	6th–3rd c. BC
Bos taurus	mandibula	47.0	82.0		35.5				Pantanello Sanctuary	6th–3rd c. BC
Bos taurus	mandibula	50.0	85.0		38.0				Pantanello Sanctuary	6th–3rd c. BC
Bos taurus	mandibula		75.0		37.0				Pantanello Necropoleis	6th–3rd c. BC
Bos taurus	mandibula		83.0		36.5				Pantanello Sanctuary	6th–3rd c. BC
Bos taurus	mandibula		84.0		37.0				Pantanello Sanctuary	6th–3rd c. BC
Bos taurus	mandibula		85.0		38.0				Incoronata	8th–6th c. BC
Bos taurus	mandibula		91.0		40.0				Pantanello Pits	Late Neolithic
Bos taurus	mandibula		93.5		40.0				Pantanello Kiln Deposit	2nd c. BC–1st c. AD
Bos taurus	mandibula				35.0				Incoronata	8th–6th c. BC
Bos taurus	mandibula				38.0				Pantanello Pits	Late Neolithic
Bos taurus	mandibula				40.0				Pantanello Kiln Deposit	2nd c. BC–1st c. AD
Bos taurus	mandibula				40.0				Pantanello Sanctuary	6th–3rd c. BC
Bos taurus	mandibula				41.0				Pantanello Kiln Deposit	2nd c. BC–1st c. AD
	Skeletal part		**LAPa**	**BFcr**		**BFcd**	**Height Fcd**		**Site name**	**Period**
Bos taurus	axis		53.0	88.0		42.0	41.0		Pantanello Sanctuary	6th–3rd c. BC

	Skeletal part			SLC	BG		GLP		Site name	Period
Bos taurus	scapula			48.5	54.0		71.0		Pantanello Sanctuary	6th–3rd c. BC
Bos taurus	scapula			50.0	55.0		75.0		Pantanello Sanctuary	6th–3rd c. BC
Bos taurus	scapula			54.0	57.5		74.0		Pantanello Kiln Deposit	2nd c. BC–1st c. AD
	Skeletal part	**GL**	**Bp**	**Dp**	**SD**	**DD**	**Bd**	**Dd**	**Site name**	**Period**
Bos taurus	humerus				36.0	44.0	90.0		Pantanello Kiln Deposit	2nd c. BC–1st c. AD
Bos taurus	humerus				39.0	44.0	94.0		Pantanello Kiln Deposit	2nd c. BC–1st c. AD
Bos taurus	radius	280.0	83.0	44.0	44.0	21.0	70.0	45.0	Pantanello Sanctuary	6th–3rd c. BC
Bos taurus	radius	282.0	80.0	40.0	37.0	20.5	69.0	43.0	Incoronata	8th–6th c. BC
Bos taurus	radius	296.0		45.0	44.0	22.0			Pantanello Kiln Deposit	2nd c. BC–1st c. AD
Bos taurus	radius	302.0	84.0	47.0	49.0	25.5		45.0	Pantanello Kiln Deposit	2nd c. BC–1st c. AD
Bos taurus	radius	321.0			49.0		88.0	53.0	Sant' Angelo Grieco	6th c. BC–1st c. AD
Bos taurus	radius	328.0	90.0	46.0	48.0	25.0		52.0	Pantanello Kiln Deposit	2nd c. BC–1st c. AD
Bos taurus	radius	334.0	98.0	53.0	42.5	26.0	82.0	52.0	Pantanello Kiln Deposit	2nd c. BC – 1st c. AD
Bos taurus	radius		75.5	28.5		32.0			Termitito	Late Bronze Age
Bos taurus	radius		82.0	47.0					Pantanello Kiln Deposit	2nd c. BC–1st c. AD
Bos taurus	radius		85.0	47.0		24.5			Pantanello Kiln Deposit	2nd c. BC–1st c. AD
Bos taurus	radius		93.0	48.0					Pantanello Sanctuary	6th–3rd c. BC
Bos taurus	radius		95.0	52.0					Pantanello Kiln Deposit	2nd c. BC–1st c. AD
Bos taurus	radius						55.0	33.0	Termitito	Late Bronze Age
Bos taurus	metacarpus	184.0	50.5	33.0	27.0	18.0	55.0	29.0	Incoronata	8th–6th c. BC
Bos taurus	metacarpus	188.0	56.0	35.0	31.0	19.5	61.0	31.5	Incoronata	8th–6th c. BC
Bos taurus	metacarpus	190.0			31.0	21.0	56.5		Pantanello Sanctuary	6th–3rd c. BC
Bos taurus	metacarpus	194.0		36.0		20.5	60.0	31.0	Pantanello Sanctuary	6th–3rd c. BC
Bos taurus	metacarpus	198.0				21.5	57.0	31.0	Pantanello Sanctuary	6th–3rd c. BC

	Skeletal part	GL	Bp	Dp	SD	DD	Bd	Dd	Site name	Period
Bos taurus	metacarpus	207.0	70.0	44.0	45.0	23.0		37.0	Pantanello Sanctuary	6th–3rd c. BC
Bos taurus	metacarpus	210.0	62.0	38.0	37.0				Pantanello Sanctuary	6th–3rd c. BC
Bos taurus	metacarpus	214.5	65.0	42.5	41.0	23.0	72.5	38.0	Sant' Angelo Grieco	6th c. BC–1st c. AD
Bos taurus	metacarpus	218.0	61.0	38.5	35.0	25.0	63.0	35.0	Pantanello Kiln Deposit	2nd c. BC–1st c. AD
Bos taurus	metacarpus	219.0	63.0	39.0	35.0	25.0	63.0	35.0	Pantanello Kiln Deposit	2nd c. BC–1st c. AD
Bos taurus	metacarpus	220.0				26.5	72.0	40.0	Pantanello Kiln Deposit	2nd c. BC–1st c. AD
Bos taurus	metacarpus	222.0	67.0	40.5	35.0	24.0	70.5	35.0	Pantanello Kiln Deposit	2nd c. BC–1st c. AD
Bos taurus	metacarpus		47.0	29.0	27.5				Incoronata	8th–6th c. BC
Bos taurus	metacarpus		51.0	32.0					Incoronata	8th–6th c. BC
Bos taurus	metacarpus		54.5	35.0	30.0				Pantanello Sanctuary	6th–3rd c. BC
Bos taurus	metacarpus		57.5	35.0					Incoronata	8th–6th c. BC
Bos taurus	metacarpus		60.0	35.0					Pantanello Sanctuary	6th–3rd c. BC
Bos taurus	metacarpus		60.0	37.5	34.0				Pantanello Sanctuary	6th–3rd c. BC
Bos taurus	metacarpus		60.0	38.0	31.0				Pantanello Sanctuary	6th–3rd c. BC
Bos taurus	metacarpus		61.0	38.5	36.0				Pantanello Sanctuary	6th–3rd c. BC
Bos taurus	metacarpus		61.0	39.0	36.5	25.0			Sant' Angelo Grieco	6th c. BC–1st c. AD
Bos taurus	metacarpus		64.0	39.0					Pantanello Sanctuary	6th–3rd c. BC
Bos taurus	metacarpus		64.0	42.0	38.5	24.5			Pantanello Kiln Deposit	2nd c. BC–1st c. AD
Bos taurus	metacarpus		66.0	40.0	40.0				Pantanello Kiln Deposit	2nd c. BC–1st c. AD
Bos taurus	metacarpus		67.0	43.0					Pantanello Sanctuary	6th–3rd c. BC
Bos taurus	metacarpus		68.0	40.0	38.0				Pantanello Kiln Deposit	2nd c. BC–1st c. AD
Bos taurus	metacarpus					19.0	52.0	28.0	Incoronata	8th–6th c. BC
Bos taurus	metacarpus					20.0	59.5	30.0	Incoronata	8th–6th c. BC
Bos taurus	metacarpus						70.0	39.0	Pantanello Kiln Deposit	2nd c. BC–1st c. AD

	Skeletal part	GL	Bp	Dp	SD	DD	Bd	Dd	Site name	Period
Bos taurus	metacarpus						72.0	38.0	Pantanello Sanctuary	6th–3rd c. BC
Bos taurus	metacarpus						76.0	42.0	Pantanello Kiln Deposit	2nd c. BC–1st c. AD
Bos taurus	femur						118.0	140.0	Pantanello Kiln Deposit	2nd c. BC–1st c. AD
Bos taurus	tibia				39.0	28.5	61.0	44.0	Sant' Angelo Grieco	6th c. BC–1st c. AD
Bos taurus	tibia						51.0	37.0	Termitito	Late Bronze Age
Bos taurus	tibia						54.0	43.0	Incoronata	8th–6th c. BC
Bos taurus	tibia						60.0	45.0	Incoronata	8th–6th c. BC
Bos taurus	tibia						60.0	45.5	Pantanello Sanctuary	6th–3rd c. BC
Bos taurus	tibia						61.0	45.5	Pantanello Sanctuary	6th–3rd c. BC
Bos taurus	tibia						61.0	42.0	Pantanello Sanctuary	6th–3rd c. BC
Bos taurus	tibia						62.5	45.5	Pantanello Sanctuary	6th–3rd c. BC
Bos taurus	tibia						68.0	52.0	Pantanello Sanctuary	6th–3rd c. BC
Bos taurus	tibia						68.5	51.0	Pantanello Necropoleis	6th–3rd c. BC
Bos taurus	tibia						69.0	51.0	Pantanello Kiln Deposit	2nd c. BC–1st c. AD
Bos taurus	tibia						70.0	52.0	Pantanello Sanctuary	6th–3rd c. BC
Bos taurus	tibia						70.0	55.0	Pantanello Sanctuary	6th–3rd c. BC
Bos taurus	tibia						70.0	55.0	Pantanello Sanctuary	6th–3rd c. BC
Bos taurus	tibia						70.5	50.0	Pantanello Sanctuary	6th–3rd c. BC
Bos taurus	tibia						71.0	50.0	Pantanello Kiln Deposit	2nd c. BC–1st c. AD
Bos taurus	tibia						73.0	53.0	Pantanello Sanctuary	6th–3rd c. BC
Bos taurus	tibia						73.0	54.0	Pantanello Kiln Deposit	2nd c. BC–1st c. AD
Bos taurus	tibia						73.0	57.0	Pantanello Sanctuary	6th–3rd c. BC
Bos taurus	tibia						73.0	67.0	Pantanello Sanctuary	6th–3rd c. BC
Bos taurus	metatarsus	212.0	46.5	44.0	24.0		53.0	31.0	Incoronata	8th–6th c. BC

	Skeletal part	GL	Bp	Dp	SD	DD	Bd	Dd	Site name	Period
Bos taurus	metatarsus	220.0							Pantanello Kiln Deposit	2nd c. BC–1st c. AD
Bos taurus	metatarsus	228.0	48.0	46.0	27.0		54.0	33.0	Pantanello Kiln Deposit	2nd c. BC–1st c. AD
Bos taurus	metatarsus	233.0	52.5	53.0	37.0	26.5	57.0	33.0	Pantanello Sanctuary	6th–3rd c. BC
Bos taurus	metatarsus	234.5	52.0	53.0	27.0	26.0	57.5	33.0	Pantanello Sanctuary	6th–3rd c. BC
Bos taurus	metatarsus	236.0	58.0	55.0	32.0	28.5	65.0	37.0	Pantanello Kiln Deposit	2nd c. BC–1st c. AD
Bos taurus	metatarsus	245.0	56.5	55.0	31.0	28.0	62.5	35.0	Pantanello Kiln Deposit	2nd c. BC–1st c. AD
Bos taurus	metatarsus	246.0	53.0	51.5	35.0	30.0			Pantanello Sanctuary	6th–3rd c. BC
Bos taurus	metatarsus	250.0	55.0	49.0	31.0		59.0	35.0	Pantanello Kiln Deposit	2nd c. BC–1st c. AD
Bos taurus	metatarsus	253.0	52.0	50.0	33.0	27.0	58.0	35.0	Pantanello Kiln Deposit	2nd c. BC–1st c. AD
Bos taurus	metatarsus	253.0	52.0	49.0	33.0	29.0	61.0	33.0	Pantanello Sanctuary	6th–3rd c. BC
Bos taurus	metatarsus	257.0	55.0	52.5	30.5	29.0			Pantanello Kiln Deposit	2nd c. BC–1st c. AD
Bos taurus	metatarsus		41.0	43.0	23.0				Incoronata	8th–6th c. BC
Bos taurus	metatarsus		42.0	42.5					Termitito	Late Bronze Age
Bos taurus	metatarsus		50.0	48.0					Pantanello Sanctuary	6th–3rd c. BC
Bos taurus	metatarsus		50.0	48.0	26.0				Pantanello Sanctuary	6th–3rd c. BC
Bos taurus	metatarsus		51.0	48.0	29.0				Pantanello Sanctuary	6th–3rd c. BC
Bos taurus	metatarsus		51.5	53.0	27.0				Pantanello Sanctuary	6th–3rd c. BC
Bos taurus	metatarsus		52.5	52.5	27.5				Pantanello Sanctuary	6th–3rd c. BC
Bos taurus	metatarsus		55.0	55.0	31.5				Pantanello Kiln Deposit	2nd c. BC–1st c. AD
Bos taurus	metatarsus						59.0	34.0	Pantanello Kiln Deposit	2nd c. BC–1st c. AD
Bos taurus	metatarsus						60.0	32.0	Sant' Angelo Grieco	6th c. BC–1st c. AD
Bos taurus	metatarsus						64.0	38.0	Pantanello Kiln Deposit	2nd c. BC–1st c. AD
Bos taurus	metatarsus						65.0	37.0	Pantanello Kiln Deposit	2nd c. BC–1st c. AD

	Skeletal part	GL	Bd		Dm				Site name	Period
Bos taurus	astragalus	59.0	41.5		33.0				Incoronata	8th–6th c. BC
Bos taurus	astragalus	61.0	40.0						Termitito	Late Bronze Age
Bos taurus	astragalus	61.0	40.0		33.5				Incoronata	8th–6th c. BC
Bos taurus	astragalus	63.0	43.0		37.0				Pantanello Pits	Late Neolithic
Bos taurus	astragalus	63.5	44.0		35.0				Incoronata	8th–6th c. BC
Bos taurus	astragalus	64.0	46.0		36.0				Pantanello Pits	Late Neolithic
Bos taurus	astragalus	65.0	41.0		34.0				Incoronata	8th–6th c. BC
Bos taurus	astragalus	66.0	48.0		39.0				Pantanello Sanctuary	6th–3rd c. BC
Bos taurus	astragalus	67.0	42.0		39.0				Pantanello Sanctuary	6th–3rd c. BC
Bos taurus	astragalus	68.0	46.0						Pantanello Sanctuary	6th–3rd c. BC
Bos taurus	astragalus	68.0	46.0		40.0				Sant' Angelo Grieco	6th c. BC–1st c. AD
Bos taurus	astragalus	68.0	48.0						Sant' Angelo Grieco	6th c. BC–1st c. AD
Bos taurus	astragalus	69.0	47.5		36.5				Sant' Angelo Grieco	6th c. BC–1st c. AD
Bos taurus	astragalus	69.0	48.5		38.0				Pantanello Sanctuary	6th–3rd c. BC
Bos taurus	astragalus	69.5	48.0		40.0				Sant' Angelo Grieco	6th c. BC–1st c. AD
Bos taurus	astragalus	70.0	46.0		38.5				Pantanello Kiln Deposit	2nd c. BC–1st c. AD
Bos taurus	astragalus	70.0	52.0		41.0				Pantanello Sanctuary	6th–3rd c. BC
Bos taurus	astragalus	70.5	47.0		38.0				Pantanello Sanctuary	6th–3rd c. BC
Bos taurus	astragalus	71.0	47.0						Pantanello Kiln Deposit	2nd c. BC–1st c. AD
Bos taurus	astragalus	71.0	49.0		40.5				San Biagio	3rd–4th c. AD
Bos taurus	astragalus	72.0	48.5						Pantanello Sanctuary	6th–3rd c. BC
Bos taurus	astragalus	72.0	49.0		42.0				Sant' Angelo Grieco	6th c. BC–1st c. AD
Bos taurus	astragalus	72.0	49.5		42.0				Pantanello Sanctuary	6th–3rd c. BC
Bos taurus	astragalus	72.0	59.5		40.0				Pantanello Sanctuary	6th–3rd c. BC

	Skeletal part	GL	Bd		Dm				Site name	Period
Bos taurus	astragalus	72.5	51.0		41.0				Pantanello Kiln Deposit	2nd c. BC–1st c. AD
Bos taurus	astragalus	73.0	53.0		41.0				Pantanello Sanctuary	6th–3rd c. BC
Bos taurus	astragalus	73.5	49.0						Pantanello Kiln Deposit	2nd c. BC–1st c. AD
Bos taurus	astragalus	74.5	49.0		43.0				Pantanello Kiln Deposit	2nd c. BC–1st c. AD
Bos taurus	astragalus	74.5	49.5		53.0				Sant' Angelo Grieco	6th c. BC–1st c. AD
Bos taurus	astragalus	74.5	53.5		44.0				Pantanello Sanctuary	6th–3rd c. BC
Bos taurus	astragalus	75.0	48.0		42.0				Pantanello Kiln Deposit	2nd c. BC–1st c. AD
Bos taurus	astragalus	77.0	53.0		40.0				Pantanello Sanctuary	6th–3rd c. BC
Bos taurus	astragalus	77.0	54.0		45.0				Pantanello Kiln Deposit	2nd c. BC–1st c. AD
Bos taurus	astragalus	77.0	55.0		42.0				Pantanello Sanctuary	6th–3rd c. BC
Bos taurus	astragalus	77.5	50.0		42.5				Pantanello Sanctuary	6th–3rd c. BC
Bos taurus	astragalus	78.0	53.0						Pantanello Sanctuary	6th–3rd c. BC
	Skeletal part	**GL**	**GB**		**Greatest depth**				**Site name**	**Period**
Bos taurus	calcaneus	127.0	44.0		49.0				Incoronata	8th–6th c. BC
Bos taurus	calcaneus	140.0	46.0		55.0				Pantanello Sanctuary	6th–3rd c. BC
Bos taurus	calcaneus	141.0	46.0		56.0				Pantanello Sanctuary	6th–3rd c. BC
Bos taurus	calcaneus	150.0	53.0		61.0				Pantanello Kiln Deposit	2nd c. BC–1st c. AD
Bos taurus	calcaneus		50.0		67.0				Pantanello Sanctuary	6th–3rd c. BC
Species	**Skeletal part**	**GL**	**Greatest d**		**Smallest d**		**Base circum.**		**Site name**	**Period**
Ovis aries	horn core	16.0	13.0		11.0				Pantanello Kiln Deposit	2nd c. BC–1st c. AD
Ovis aries	horn core	58.0			15.0				Incoronata	8th–6th c. BC
Ovis aries	horn core	63.0	27.0		17.0		75.0		Incoronata	8th–6th c. BC
Ovis aries	horn core	210.0	62.0		48.0		176.0		Pantanello Pits	Late Neolithic
Ovis aries	horn core		49.0		33.0				Pantanello Kiln Deposit	2nd c. BC–1st c. AD

Species	Skeletal part	GL	Greatest d		Smallest d		Base circum.		Site name	Period
Ovis aries	horn core		53.5		38.5		150.0		Sant' Angelo Grieco	6th c. BC–1st c. AD
Ovis aries	horn core		60.0		51.0				Pantanello Kiln Deposit	2nd c. BC–1st c. AD
Ovis aries	horn core		61.0		48.0				Pantanello Pits	Late Neolithic
	Skeletal part		**LAd**	**BFcr**	**BFcd**		**H**		**Site name**	**Period**
Ovis aries	atlas		20.0	59.0	46.0		43.0		Pantanello Pits	Late Neolithic
	Skeletal part			**SLC**	**BG**		**GLP**		**Site name**	**Period**
Ovis aries	scapula			18.5	18.0		28.8		Incoronata	8th–6th c. BC
Ovis aries	scapula			19.0	19.0		34.0		Incoronata	8th–6th c. BC
Ovis aries	scapula			21.0	17.7		33.0		Pantanello Pits	Late Neolithic
Ovis aries	scapula				19.0		31.5		Termitito	Late Bronze Age
	Skeletal part	**GL**	**Bp**	**Dp**	**SD**	**DD**	**Bd**	**Dd**	**Site name**	**Period**
Ovis aries	humerus				13.0	14.0	27.0	23.0	Termitito	Late Bronze Age
Ovis aries	humerus				16.0	18.0	32.0	27.7	Incoronata	8th–6th c. BC
Ovis aries	humerus						28.0	24.5	Termitito	Late Bronze Age
Ovis aries	humerus						29.3	25.0	Incoronata	8th–6th c. BC
Ovis aries	humerus						30.0	27.0	Incoronata	8th–6th c. BC
Ovis aries	humerus						30.5		Pantanello Kiln Deposit	2nd c. BC–1st c. AD
Ovis aries	humerus						32.5		Pantanello Kiln Deposit	2nd c. BC–1st c. AD
Ovis aries	humerus						33.0	27.8	Incoronata	8th–6th c. BC
Ovis aries	humerus						35.0		Pantanello Kiln Deposit	2nd c. BC–1st c. AD
Ovis aries	radius	136.5	27.0	15.0	13.0	7.0	24.0	18.0	Termitito	Late Bronze Age
Ovis aries	radius	160.0	29.0	15.5	15.0	8.0	27.5	19.5	Pantanello Pits	Late Neolithic
Ovis aries	radius		29.5	15.0					Termitito	Late Bronze Age
Ovis aries	radius		30.0	17.0					Pantanello Pits	Late Neolithic

	Skeletal part	GL	Bp	Dp	SD	DD	Bd	Dd	Site name	Period
Ovis aries	radius		31.0	17.0					Pantanello Pits	Late Neolithic
Ovis aries	radius		31.0	17.3	15.0	7.5			Pantanello Sanctuary	6th–3rd c. BC
Ovis aries	radius		32.0	16.0					Pantanello Pits	Late Neolithic
Ovis aries	radius		33.0	17.0					Pantanello Kiln Deposit	2nd c. BC–1st c. AD
Ovis aries	radius		34.3	17.0					Pantanello Pits	Late Neolithic
Ovis aries	radius		36.0	19.0					Pantanello Pits	Late Neolithic
Ovis aries	metacarpus	124.0	27.0		15.0	11.0	27.0	18.0	Pantanello Kiln Deposit	2nd c. BC–1st c. AD
Ovis aries	metacarpus	129.5	24.0	17.0	13.8	10.0	25.8	17.0	Pantanello Kiln Deposit	2nd c. BC–1st c. AD
Ovis aries	metacarpus		20.0	15.0	11.8				Incoronata	8th–6th c. BC
Ovis aries	metacarpus		20.5	16.0	13.0				Pantanello Pits	Late Neolithic
Ovis aries	metacarpus		22.3	16.5	13.5				Termitito	Late Bronze Age
Ovis aries	metacarpus		23.0	17.3	13.5				Pantanello Sanctuary	6th–3rd c. BC
Ovis aries	metacarpus		25.0	19.0					San Biagio	3rd–4th c. AD
Ovis aries	metacarpus		26.0	18.0	14.0				Pantanello Kiln Deposit	2nd c. BC–1st c. AD
Ovis aries	metacarpus					8.5	22.0	14.0	Termitito	Late Bronze Age
Ovis aries	metacarpus					10.0	24.5	16.5	Pantanello Pits	Late Neolithic
Ovis aries	metacarpus					10.5	26.0	16.0	Pantanello Pits	Late Neolithic
Ovis aries	metacarpus				15.0	10.0	26.0	17.0	Pantanello Pits	Late Neolithic
Ovis aries	femur	162.0	43.0	23.0	16.0	17.0	37.0	46.0	Incoronata	8th–6th c. BC
Ovis aries	tibia		40.0	41.0					Pantanello Kiln Deposit	2nd c. BC–1st c. AD
Ovis aries	metatarsus	117.0	19.0		11.5	10.0	24.0	15.0	Incoronata	8th–6th c. BC
Ovis aries	metatarsus	126.0	18.0	19.5	10.0	9.5	23.0	15.0	Termitito	Late Bronze Age
Ovis aries	metatarsus	137.0			12.0				Incoronata	8th–6th c. BC
Ovis aries	metatarsus		20.0	21.0	11.5				Pantanello Pits	Late Neolithic

	Skeletal part	**GL**	**Bp**	**Dp**	**SD**	**DD**	**Bd**	**Dd**	**Site name**	**Period**
Ovis aries	metatarsus		21.0	21.0	11.5				Incoronata	8th–6th c. BC
Ovis aries	metatarsus		21.0	22.0	12.5				Pantanello Pits	Late Neolithic
Ovis aries	metatarsus		21.0	22.0	13.0				Pantanello Pits	Late Neolithic
Ovis aries	metatarsus		22.5	22.5	12.0				Pantanello Kiln Deposit	2nd c. BC–1st c. AD
Ovis aries	metatarsus				10.0	8.7	24.0	14.8	Incoronata	8th–6th c. BC
Ovis aries	metatarsus				10.8	9.0	23.0	15.8	Termitito	Late Bronze Age
Ovis aries	metatarsus					8.8	24.0	16.0	Incoronata	8th–6th c. BC
Ovis aries	metatarsus					9.0	23.0	15.0	Pantanello Pits	Late Neolithic
Ovis aries	metatarsus					10.5	24.5	17.5	Incoronata	8th–6th c. BC
Ovis aries	metatarsus						23.8	16.5	Pantanello Pits	Late Neolithic
Ovis aries	metatarsus						25.5	18.0	Pantanello Pits	Late Neolithic
Ovis aries	metatarsus						26.0	18.0	Pantanello Pits	Late Neolithic
Ovis aries	metatarsus						26.0	18.0	Pantanello Pits	Late Neolithic
	Skeletal part	**GL**	**Bd**		**Dm**				**Site name**	**Period**
Ovis aries	astragalus	26.0	18.0		15.0				Termitito	Late Bronze Age
Ovis aries	astragalus	26.0	19.0		15.5				Termitito	Late Bronze Age
Ovis aries	astragalus	29.0	20.5		17.0				Pantanello Pits	Late Neolithic
Ovis aries	astragalus	32.0	22.5		18.5				Pantanello Sanctuary	6th–3rd c. BC
Ovis aries	astragalus	32.0	23.0		18.5				San Biagio	3rd–4th c. AD
	Skeletal part	**GL**	**GB**		**Greatest depth**				**Site name**	**Period**
Ovis aries	calcaneus	56.5	21.5		24.0				Pantanello Kiln Deposit	2nd c. BC–1st c. AD
Ovis aries	calcaneus	59.0	21.5		25.0				Pantanello Pits	Late Neolithic
Species	**Skeletal part**	**GL**	**Greatest d**		**Smallest d**		**Base circum.**		**Site name**	**Period**
Capra hircus	horn core	142.0	30.0		21.0		82.0		Pantanello Sanctuary	6th–3rd c. BC

Species	**Skeletal part**	**GL**	**Greatest d**		**Smallest d**		**Base circum.**		**Site name**	**Period**
Capra hircus	horn core		28.0		18.0		75.0		Pantanello Sanctuary	6th–3rd c. BC
Capra hircus	horn core		32.0		20.5		84.0		Incoronata	8th–6th c. BC
Capra hircus	horn core		33.0		21.5		89.0		Incoronata	8th–6th c. BC
Capra hircus	horn core		39.0		26.0		106.0		Pantanello Sanctuary	6th–3rd c. BC
	Skeletal part	**GL**	**Bp**	**Dp**	**SD**	**DD**	**Bd**	**Dd**	**Site name**	**Period**
Capra hircus	humerus						30.0	24.0	Incoronata	8th–6th c. BC
Capra hircus	metacarpus		25.0	16.0	14.7				Incoronata	8th–6th c. BC
Capra hircus	metacarpus		25.0	17.5					Pantanello Pits	Late Neolithic
Capra hircus	metacarpus					10.3	26.5	16.0	Incoronata	8th–6th c. BC
Capra hircus	metacarpus						28.0	17.0	Incoronata	8th–6th c. BC
Capra hircus	metatarsus	126.0	20.0	19.0	13.5	10.7	25.0	16.2	Termitito	Late Bronze Age
Capra hircus	metatarsus		18.0	16.5					Pantanello Pits	Late Neolithic
Capra hircus	radius		26.0	14.5	14.5	8.0			Incoronata	8th–6th c. BC
Capra hircus	radius		28.5	15.0	17.0	9.0			Pantanello Sanctuary	6th–3rd c. BC
Capra hircus	radius		32.0	16.5					San Biagio	3rd–4th c. AD
Capra hircus	femur						37.0	43.0	Incoronata	8th–6th c. BC
Species	**Skeletal part**	**P2–P4 length**	**M1–M3 length**		**M3 length**				**Site name**	**Period**
Equus caballus	maxilla	81.0	72.0						Pantanello Necropoleis	6th–3rd c. BC
Equus caballus	mandibula	82.0	79.0		33.0				Pantanello Necropoleis	6th–3rd c. BC
	Skeletal part		**LAPa**	**BFcr**	**L of dens**	**B of dens**			**Site name**	**Period**
Equus caballus	axis		60.0	80.0	25.0	35.0			Pantanello Sanctuary	6th–3rd c. BC
	Skeletal part	**GL**	**Bp**	**Dp**	**SD**	**DD**	**Bd**	**Dd**	**Site name**	**Period**
Equus caballus	humerus				53.0	39.0	77.0		Pantanello Kiln Deposit	2nd c. BC–1st c. AD
Equus caballus	radius	330.0			36.0	26.0			Pantanello Sanctuary	6th–3rd c. BC

	Skeletal part	GL	Bp	Dp	SD	DD	Bd	Dd	Site name	Period
Equus caballus	radius	340.0		49.0	40.0	28.0	77.0	47.0	Pantanello Kiln Deposit	2nd c. BC–1st c. AD
Equus caballus	radius		77.0	42.5					Sant' Angelo Grieco	6th c. BC–1st c. AD
Equus caballus	radius		80.0	43.0					Pantanello Sanctuary	6th–3rd c. BC
Equus caballus	radius		83.0	50.0	40.0	29.0			Pantanello Necropoleis	6th–3rd c. BC
Equus caballus	radius				35.0	23.5	72.0	45.0	Pantanello Kiln Deposit	2nd c. BC–1st c. AD
Equus caballus	radius				38.0		78.0	46.0	Pantanello Kiln Deposit	2nd c. BC–1st c. AD
Equus caballus	radius						74.0	42.0	Pantanello Kiln Deposit	2nd c. BC–1st c. AD
Equus caballus	metacarpus	209.0	45.0	30.0	31.0	20.0	43.0	32.5	Pantanello Kiln Deposit	2nd c. BC–1st c. AD
Equus caballus	metacarpus	221.0	46.0	34.0	31.5	20.5	46.0	33.0	Pantanello Sanctuary	6th–3rd c. BC
Equus caballus	metacarpus	223.0	48.0		32.0	22.5	49.5	35.0	Pantanello Sanctuary	6th–3rd c. BC
Equus caballus	metacarpus	223.0	50.0	35.0	34.0	24.0	52.0	37.0	Pantanello Sanctuary	6th–3rd c. BC
Equus caballus	metacarpus	226.0	50.0	36.0	33.0	22.0	49.5	36.5	Pantanello Kiln Deposit	2nd c. BC–1st c. AD
Equus caballus	metacarpus	227.0	51.0	35.0	34.0		50.0	36.0	Pantanello Kiln Deposit	2nd c. BC–1st c. AD
Equus caballus	metacarpus	240.0	50.0	35.5	35.0	24.5	49.0	38.0	Pantanello Necropoleis	6th–3rd c. BC
Equus caballus	metacarpus		52.0	38.0	33.5				Pantanello Sanctuary	6th–3rd c. BC
Equus caballus	metacarpus					21.0	46.0	25.5	Pantanello Sanctuary	6th–3rd c. BC
Equus caballus	phalanx proximalis anterior	75.4	49.3	29.8	32.5	19.6	43.6	24.1	Pantanello Sanctuary	6th–3rd c. BC
Equus caballus	phalanx proximalis anterior	77.0	52.0	35.0	33.0	18.5	42.5	25.0	Pantanello Sanctuary	6th–3rd c. BC
Equus caballus	phalanx proximalis anterior	81.0	57.5	41.0	35.5	21.0	45.0	26.5	Pantanello Necropoleis	6th–3rd c. BC
Equus caballus	phalanx medialis anterior	42.5	54.5	33.5	44.0	23.0	48.0		Pantanello Kiln Deposit	2nd c. BC–1st c. AD
Equus caballus	phalanx medialis anterior	46.0	55.0	33.0	46.5	26.0	52.0	29.0	Pantanello Kiln Deposit	2nd c. BC–1st c. AD
Equus caballus	tibia	362.0			40.0	30.0	72.0	48.0	Pantanello Sanctuary	6th–3rd c. BC
Equus caballus	tibia	369.0			41.0	31.0		47.0	Pantanello Necropoleis	6th–3rd c. BC

	Skeletal part	GL	Bp	Dp	SD	DD	Bd	Dd	Site name	Period
Equus caballus	metatarsus	263.0			29.0	23.5	50.0	37.0	Pantanello Kiln Deposit	2nd c. BC–1st c. AD
Equus caballus	metatarsus	266.0							Pantanello Kiln Deposit	2nd c. BC–1st c. AD
Equus caballus	metatarsus	270.0			29.5	26.0	52.0	40.0	Pantanello Kiln Deposit	2nd c. BC–1st c. AD
Equus caballus	metatarsus	270.0			32.5	27.0	51.0	37.0	Pantanello Sanctuary	6th–3rd c. BC
Equus caballus	metatarsus	274.0	50.0	48.0	31.0	26.0	51.0	40.0	Pantanello Kiln Deposit	2nd c. BC–1st c. AD
Equus caballus	metatarsus	286.0	52.0	47.0	33.5	27.0	51.0	39.5	Pantanello Necropoleis	6th–3rd c. BC
Equus caballus	metatarsus		54.0	51.0	33.0				Pantanello Kiln Deposit	2nd c. BC–1st c. AD
Equus caballus	metatarsus				29.0				Pantanello Kiln Deposit	2nd c. BC–1st c. AD
Equus caballus	metatarsus				27.0	24.0	48.0	37.0	Pantanello Sanctuary	6th–3rd c. BC
	Skeletal part	**GL**	**Bd**		**Dm**				**Site name**	**Period**
Equus caballus	astragalus	62.0	60.0						Pantanello Kiln Deposit	2nd c. BC–1st c. AD
Equus caballus	astragalus	62.0	61.5		57.0				Pantanello Sanctuary	6th–3rd c. BC
Equus caballus	astragalus	64.0	59.0		60.0				Pantanello Sanctuary	6th–3rd c. BC
Equus caballus	astragalus	64.0	60.0		58.0				Pantanello Kiln Deposit	2nd c. BC–1st c. AD
Equus caballus	astragalus	64.0	61.0						Pantanello Necropoleis	6th–3rd c. BC
Equus caballus	astragalus	65.0	62.0		63.0				Pantanello Kiln Deposit	2nd c. BC–1st c. AD
Equus caballus	astragalus	65.0	63.0		60.0				Pantanello Sanctuary	6th–3rd c. BC
	Skeletal part	**GL**	**GB**		**Greatest depth**				**Site name**	**Period**
Equus caballus	calcaneus	113.0							Pantanello Kiln Deposit	2nd c. BC–1st c. AD
Equus caballus	calcaneus	114.0	55.0		57.0				Pantanello Necropoleis	6th–3rd c. BC
	Skeletal part	**GL**	**Bp**	**Dp**	**SD**	**DD**	**Bd**	**Dd**	**Site name**	**Period**
Equus caballus	phalanx proximalis post.	73.0	54.0	36.0	33.0	18.5	44.0	24.0	Pantanello Kiln Deposit	2nd c. BC–1st c. AD
Equus caballus	phalanx proximalis post.	73.0	57.0	41.0	65.0	20.0	43.0	26.5	Pantanello Kiln Deposit	2nd c. BC–1st c. AD
Equus caballus	phalanx proximalis post.	75.5	53.0	38.0	29.0	18.0	40.0	25.0	Pantanello Kiln Deposit	2nd c. BC–1st c. AD

	Skeletal part	GL	Bp	Dp	SD	DD	Bd	Dd	Site name	Period
Equus caballus	phalanx proximalis post.	77.0		40.0	32.0	20.0	44.0	25.5	Pantanello Sanctuary	6th–3rd c. BC
Equus caballus	phalanx proximalis post.	77.5	53.0	38.0	31.0	20.0	40.5	25.5	Incoronata	8th–6th c. BC
Equus caballus	phalanx media post.	42.0	55.0	33.0	47.5	25.5	53.0	29.0	Pantanello Kiln Deposit	2nd c. BC–1st c. AD
Equus caballus	phalanx media post.	42.5	51.0	32.0	41.0	23.0	45.5	28.0	Pantanello Sanctuary	6th–3rd c. BC
Species	**Skeletal part**	**GL**	**Bd**		**Dm**				**Site name**	**Period**
Asinus asinus	astragalus	48.0	46.0		48.0				Pantanello Kiln Deposit	2nd c. BC–1st c. AD
	Skeletal part	**GL**	**Bp**	**Dp**	**SD**	**DD**	**Bd**	**Dd**	**Site name**	**Period**
Asinus asinus	phalanx proximalis ant.	63.5	35.0	27.0	22.0	13.7	32.0	19.0	Pantanello Sanctuary	6th–3rd c. BC
Species	**Skeletal part**	**P2–P4 length**	**M1–M3 length**		**M3 length**				**Site name**	**Period**
Sus domesticus	maxilla	42.5							Incoronata	8th–6th c. BC
Sus domesticus	maxilla		62.0		28.5				Pantanello Sanctuary	6th–3rd c. BC
Sus domesticus	maxilla				27.0				Incoronata	8th–6th c. BC
Sus domesticus	maxilla				28.0				Incoronata	8th–6th c. BC
Sus domesticus	maxilla				29.0				Incoronata	8th–6th c. BC
Sus domesticus	maxilla				31.5				Termitito	Late Bronze Age
Sus domesticus	maxilla				31.5				Pantanello Sanctuary	6th–3rd c. BC
Sus domesticus	maxilla				33.0				Incoronata	8th–6th c. BC
Sus domesticus	mandibula	45.0							Sant' Angelo Grieco	6th c. BC–1st c. AD
Sus domesticus	mandibula	47.5							Incoronata	8th–6th c. BC
Sus domesticus	mandibula	48.0							Incoronata	8th–6th c. BC
Sus domesticus	mandibula	50.0							Pantanello Kiln Deposit	2nd c. BC–1st c. AD
Sus domesticus	mandibula	51.0							Incoronata	8th–6th c. BC
Sus domesticus	mandibula	54.0							Pantanello Kiln Deposit	2nd c. BC–1st c. AD
Sus domesticus	mandibula	54.0							Termitito	Late Bronze Age

	Skeletal part	P2–P4 length	M1–M3 length		M3 length				Site name	Period
Sus domesticus	mandibula		61.0		31.0				Pantanello Sanctuary	6th–3rd c. BC
Sus domesticus	mandibula				28.0				Incoronata	8th–6th c. BC
Sus domesticus	mandibula				28.0				Incoronata	8th–6th c. BC
Sus domesticus	mandibula				30.0				Pantanello Sanctuary	6th–3rd c. BC
Sus domesticus	mandibula				30.5				Incoronata	8th–6th c. BC
Sus domesticus	mandibula				31.0				Incoronata	8th–6th c. BC
Sus domesticus	mandibula				31.0				Incoronata	8th–6th c. BC
Sus domesticus	mandibula				32.0				Incoronata	8th–6th c. BC
Sus domesticus	mandibula				32.0				Incoronata	8th–6th c. BC
Sus domesticus	mandibula				32.0				Incoronata	8th–6th c. BC
Sus domesticus	mandibula				33.0				Incoronata	8th–6th c. BC
Sus domesticus	mandibula				34.5				Incoronata	8th–6th c. BC
Sus domesticus	mandibula				35.0				Incoronata	8th–6th c. BC
Sus domesticus	mandibula				37.0				Incoronata	8th–6th c. BC
	Skeletal part	**LCDe**	**LAd**	**BFcr**	**BFcd**		**H**		**Site name**	**Period**
Sus domesticus	atlas	15.0	14.0	48.0	39.5		42.0		Sant' Angelo Grieco	6th c. BC–1st c. AD
	Skeletal part		**LAPa**	**BFcr**	**L of dens**	**B of dens**			**Site name**	**Period**
Sus domesticus	axis		13.5	44.0	13.0	10.0			Termitito	Late Bronze Age
	Skeletal part			**SLC**	**BG**		**GLP**		**Site name**	**Period**
Sus domesticus	scapula			23.5	21.5		32.0		Incoronata	8th–6th c. BC
Sus domesticus	scapula				21.3		34.5		Termitito	Late Bronze Age
Sus domesticus	scapula				22.0		31.0		Pantanello Kiln Deposit	2nd c. BC–1st c. AD
Sus domesticus	scapula				26.0		37.5		San Biagio	3rd–4th c. AD
Sus domesticus	humerus						37.0	39.0	Incoronata	8th–6th c. BC

	Skeletal part	GL	Bp	Dp	SD	DD	Bd	Dd	Site name	Period
Sus domesticus	humerus						38.0		Pantanello Kiln Deposit	2nd c. BC–1st c. AD
Sus domesticus	humerus						40.5		Pantanello Kiln Deposit	2nd c. BC–1st c. AD
Sus domesticus	radius		27.0	20.0	16.0	11.0			Incoronata	8th–6th c. BC
Sus domesticus	radius		28.0	18.0					Incoronata	8th–6th c. BC
Sus domesticus	radius		28.0	19.0	17.0	11.0			Termitito	Late Bronze Age
Sus domesticus	radius		29.0	20.2	18.5	12.7			Termitito	Late Bronze Age
Sus domesticus	radius		30.0	20.0					Termitito	Late Bronze Age
Sus domesticus	radius		33.0	24.0				32.0	San Biagio	3rd–4th c. AD
Sus domesticus	radius						34.0	26.0	Pantanello Kiln Deposit	2nd c. BC–1st c. AD
Sus domesticus	tibia				17.0	12.0	28.5	24.5	Incoronata	8th–6th c. BC
Sus domesticus	tibia				20.0	14.0	29.0	25.0	Pantanello Kiln Deposit	2nd c. BC–1st c. AD
Sus domesticus	tibia						29.0	25.0	Incoronata	8th–6th c. BC
Sus domesticus	tibia						30.3	28.0	Pantanello Kiln Deposit	2nd c. BC–1st c. AD
Sus domesticus	tibia						31.0	26.5	San Biagio	3rd–4th c. AD
	Skeletal part	**GL**	**Bd**		**Dm**				**Site name**	**Period**
Sus domesticus	astragalus	35.0	23.0		20.5				Termitito	Late Bronze Age
Sus domesticus	astragalus	36.0	22.0		20.5				Incoronata	8th–6th c. BC
Sus domesticus	astragalus	36.5	21.5		21.5				Incoronata	8th–6th c. BC
Sus domesticus	astragalus	37.0	22.0		20.5				Incoronata	8th–6th c. BC
Sus domesticus	astragalus	48.0	28.0						Pantanello Pits	Late Neolithic
Species	**Skeletal part**		**M1–M3 length**		**P4 length**				**Site name**	**Period**
Canis familiaris	maxilla				17.0				Incoronata	8th–6th c. BC
Canis familiaris	maxilla				17.5				Termitito	Late Bronze Age
Canis familiaris	mandibula		30.0		18.0				Incoronata	8th–6th c. BC

	Skeletal part		M1–M3 length		M1 length				Site name	Period
Canis familiaris	mandibula		37.0		22.0				Pantanello Kiln Deposit	2nd c. BC–1st c. AD
Canis familiaris	mandibula		41.0		24.0				Pantanello Kiln Deposit	2nd c. BC–1st c. AD
Canis familiaris	mandibula				20.0				Incoronata	8th–6th c. BC
Canis familiaris	mandibula				22.0				Pantanello Kiln Deposit	2nd c. BC–1st c. AD
	Skeletal part	**Lav**	**LAd**		**BDcr**	**BDcd**	**H**		**Site name**	**Period**
Canis familiaris	atlas	8.5	17.0		38.0	27.5	31.0		Termitito	Late Bronze Age
Canis familiaris	atlas		13.0		36.0		28.0		Pantanello Sanctuary	6th–3rd c. BC
	Skeletal part	**LCDe**	**LAPa**	**BFcr**	**BFcd**	**B greatest**	**L of dens**	**B of dens**	**Site name**	**Period**
Canis familiaris	axis	53.0	22.0	33.0	12.5	20.5	8.0	4.7	Pantanello Kiln Deposit	2nd c. BC–1st c. AD
	Skeletal part	**GL**	**Bp**	**Dp**	**SD**	**DD**	**Bd**	**Dd**	**Site name**	**Period**
Canis familiaris	humerus			43.0			36.0		Pantanello Kiln Deposit	2nd c. BC–1st c. AD
Canis familiaris	humerus				12.0	12.0	34.3		Pantanello Sanctuary	6th–3rd c. BC
Canis familiaris	humerus						44.0		Pantanello Kiln Deposit	2nd c. BC–1st c. AD
Canis familiaris	radius		17.3	12.0	13.0	6.5			Incoronata	8th–6th c. BC
Canis familiaris	radius		19.0	12.5		7.0			Pantanello Sanctuary	6th–3rd c. BC
Canis familiaris	radius		20.0	13.0	12.0	3.5			Pantanello Kiln Deposit	2nd c. BC–1st c. AD
Canis familiaris	radius						27.0	17.0	Pantanello Kiln Deposit	2nd c. BC–1st c. AD
Canis familiaris	femur		38.5	20.0					Pantanello Sanctuary	6th–3rd c. BC
Canis familiaris	femur		42.0	21.0					Pantanello Pits	Late Neolithic
Canis familiaris	femur						30.0	30.5	Incoronata	8th–6th c. BC
Canis familiaris	femur						30.5	32.0	Sant' Angelo Grieco	6th c. BC–1st c. AD
Canis familiaris	tibia	190.0			12.7	10.5	22.5	16.0	Pantanello Sanctuary	6th–3rd c. BC
Canis familiaris	tibia				13.5	12.0	22.0	16.5	Incoronata	8th–6th c. BC
Canis familiaris	tibia						26.0	18.5	Pantanello Kiln Deposit	2nd c. BC–1st c. AD

	Skeletal part	GL	GB		Greatest depth				Site name	Period
Canis familiaris	calcaneus	37.0	15.0		16.0				San Biagio	3rd–4th c. AD
Canis familiaris	calcaneus	46.7	19.0		19.0				Incoronata	8th–6th c. BC
Canis familiaris	calcaneus	51.0	22.0		22.0				Pantanello Kiln Deposit	2nd c. BC–1st c. AD
Canis familiaris	calcaneus	55.0	24.5		24.5				Pantanello Kiln Deposit	2nd c. BC–1st c. AD
Species	**Skeletal part**	**GL**	**Bp**	**Dp**	**SD**	**DD**	**Bd**	**Dd**	**Site name**	**Period**
Felis catus	tibia				7.9	7.1	14.8	10.3	Pantanello Sanctuary	6th–3rd c. BC
Species	**Skeletal part**			**SLC**	**BG**		**GLP**		**Site name**	**Period**
Bos primigenius	scapula			56.0	61.0		82.0		Pantanello Kiln Deposit	2nd c. BC–1st c. AD
Bos primigenius	scapula			57.0			80.0		Pantanello Pits	Late Neolithic
Bos primigenius	scapula			58.0	66.0		86.0		Pantanello Kiln Deposit	2nd c. BC–1st c. AD
Bos primigenius	scapula			60.0	57.0		85.0		Pantanello Kiln Deposit	2nd c. BC–1st c. AD
Bos primigenius	scapula				61.0		83.0		Pantanello Sanctuary	6th–3rd c. BC
Bos primigenius	scapula				68.0				Pantanello Sanctuary	6th–3rd c. BC
	Skeletal part	**GL**	**Bp**	**Dp**	**SD**	**DD**	**Bd**	**Dd**	**Site name**	**Period**
Bos primigenius	radius		96.0	52.0					Pantanello Kiln Deposit	2nd c. BC–1st c. AD
Bos primigenius	tibia				48.0	34.0	75.0	56.0	Pantanello Sanctuary	6th–3rd c. BC
Bos primigenius	metatarsus						70.0	38.0	Pantanello Kiln Deposit	2nd c. BC–1st c. AD
Species	**Skeletal part**	**GL**	**Bp**	**Dp**	**SD**	**DD**	**Bd**	**Dd**	**Site name**	**Period**
Rupicapra rupicapra	metacarpus	149.0	25.3	18.0	16.0	13.0	28.0	19.0	Pantanello Pits	Late Neolithic
Rupicapra rupicapra	metacarpus					10.5	26.2	18.0	Pantanello Pits	Late Neolithic
Species	**Skeletal part**		**Greatest d**	**Smallest d**					**Site name**	**Period**
Cervus elaphus	antler burr		72.5	65.0					Pantanello Sanctuary	6th–3rd c. BC
Cervus elaphus	antler burr		74.1	68.7					Pantanello Sanctuary	6th–3rd c. BC
Cervus elaphus	mandibula	41.5							Pantanello Kiln Deposit	2nd c. BC–1st c. AD

	Skeletal part	P2–P4 length	M1–M3 length		M3 length				Site name	Period
Cervus elaphus	mandibula				31.0				Pantanello Kiln Deposit	2nd c. BC– 1st c. AD
	Skeletal part	**LCDe**	**LAd**	**BFcr**	**BFcd**		**H**		**Site name**	**Period**
Cervus elaphus	atlas	46.0		58.0	68.0				Pantanello Sanctuary	6th–3rd c. BC
	Skeletal part			**SLC**	**BG**		**GLP**		**Site name**	**Period**
Cervus elaphus	scapula			36.0			53.0		Pantanello Kiln Deposit	2nd c. BC– 1st c. AD
Cervus elaphus	scapula			38.0	32.0		56.0		Pantanello Kiln Deposit	2nd c. BC– 1st c. AD
Cervus elaphus	scapula			42.0	34.0		56.0		Pantanello Sanctuary	6th–3rd c. BC
Cervus elaphus	scapula			51.0	41.0		65.0		Pantanello Sanctuary	6th–3rd c. BC
	Skeletal part	**GL**	**Bp**	**Dp**	**SD**	**DD**	**Bd**	**Dd**	**Site name**	**Period**
Cervus elaphus	humerus				26.0	38.0	59.0	58.0	Pantanello Sanctuary	6th–3rd c. BC
Cervus elaphus	humerus				28.0	40.0	56.0	55.0	Pantanello Sanctuary	6th–3rd c. BC
Cervus elaphus	humerus						55.5		Incoronata	8th–6th c. BC
Cervus elaphus	humerus						57.0	56.0	Pantanello Sanctuary	6th–3rd c. BC
Cervus elaphus	radius		50.0	30.0					Incoronata	8th–6th c. BC
Cervus elaphus	radius		52.0	30.0					Pantanello Sanctuary	6th–3rd c. BC
Cervus elaphus	radius		53.5	30.5					Pantanello Sanctuary	6th–3rd c. BC
Cervus elaphus	radius		54.5	34.0	33.0	19.0			Pantanello Kiln Deposit	2nd c. BC– 1st c. AD
Cervus elaphus	radius		55.0	31.0					Pantanello Sanctuary	6th–3rd c. BC
Cervus elaphus	radius						48.0	34.0	Pantanello Kiln Deposit	2nd c. BC– 1st c. AD
Cervus elaphus	radius						50.0	37.5	Pantanello Kiln Deposit	2nd c. BC– 1st c. AD
Cervus elaphus	metacarpus					17.0	39.5	25.5	Pantanello Sanctuary	6th–3rd c. BC
	Skeletal part	**GL**	**Bp**	**Dp**	**SD**	**DD**	**Bd**	**Dd**	**Site name**	**Period**
Cervus elaphus	metacarpus					19.0	45.5	28.0	Pantanello Kiln Deposit	2nd c. BC– 1st c. AD
Cervus elaphus	metacarpus					19.5	44.5	28.5	Pantanello Kiln Deposit	2nd c. BC– 1st c. AD

	Skeletal part	GL	Bp	Dp	SD	DD	Bd	Dd	Site name	Period
Cervus elaphus	metacarpus					20.0	46.0	30.0	Pantanello Kiln Deposit	2nd c. BC–1st c. AD
Cervus elaphus	metacarpus					21.0	43.0	29.0	Pantanello Sanctuary	6th–3rd c. BC
Cervus elaphus	metacarpus						40.0	27.5	Pantanello Sanctuary	6th–3rd c. BC
Cervus elaphus	tibia				25.5	23.0	45.0	35.0	Incoronata	8th–6th c. BC
Cervus elaphus	tibia				28.0	25.0	44.0	35.0	Pantanello Sanctuary	6th–3rd c. BC
Cervus elaphus	tibia				31.5	26.0	51.0	39.5	Pantanello Sanctuary	6th–3rd c. BC
Cervus elaphus	tibia						48.0	37.5	Incoronata	8th–6th c. BC
Cervus elaphus	tibia						50.5	38.0	Pantanello Kiln Deposit	2nd c. BC–1st c. AD
Cervus elaphus	tibia						50.5	39.0	Pantanello Sanctuary	6th–3rd c. BC
Cervus elaphus	tibia						51.0	41.0	Pantanello Kiln Deposit	2nd c. BC–1st c. AD
Cervus elaphus	tibia						53.0	40.0	Sant' Angelo Grieco	6th c. BC–1st c. AD
Cervus elaphus	metatarsus		39.5	41.0					Pantanello Kiln Deposit	2nd c. BC–1st c. AD
Cervus elaphus	metatarsus					18.3	43.0	28.0	Termitito	Late Bronze Age
Cervus elaphus	metatarsus					19.0	39.5	27.0	Incoronata	8th–6th c. BC
Cervus elaphus	metatarsus					20.0	43.0	29.0	Pantanello Sanctuary	6th–3rd c. BC
Cervus elaphus	metatarsus						40.0	28.0	Pantanello Kiln Deposit	2nd c. BC–1st c. AD
	Skeletal part	**GL**	**Bd**		**Dm**				**Site name**	**Period**
Cervus elaphus	astragalus	51.0	32.0		28.5				Pantanello Kiln Deposit	2nd c. BC–1st c. AD
Cervus elaphus	astragalus	51.0	33.5		27.0				Incoronata	8th–6th c. BC
Cervus elaphus	astragalus	52.0	31.0		29.0				Incoronata	8th–6th c. BC
Cervus elaphus	astragalus	52.0	32.0		30.0				Pantanello Kiln Deposit	2nd c. BC–1st c. AD
Cervus elaphus	astragalus	53.5	36.0		30.0				Incoronata	8th–6th c. BC
Cervus elaphus	astragalus	58.5	37.0		32.5				Pantanello Sanctuary	6th–3rd c. BC
Cervus elaphus	calcaneus	124.0	34.0		45.0				Pantanello Kiln Deposit	2nd c. BC–1st c. AD

	Skeletal part	GL	Bp	Dp	SD	DD	Bd	Dd	Site name	Period
Cervus elaphus	phalanx proximalis	49.0	22.0	27.0	18.0	15.5	20.5	18.5	Incoronata	8th–6th c. BC
Cervus elaphus	phalanx proximalis	50.0	21.0	24.7	15.0	14.0	19.5	16.0	Pantanello Kiln Deposit	2nd c. BC–1st c. AD
Cervus elaphus	phalanx proximalis	50.0	21.5	25.0	17.0	15.0	20.0	18.0	Pantanello Kiln Deposit	2nd c. BC–1st c. AD
Cervus elaphus	phalanx proximalis	50.0	23.5	27.5	18.0	15.5	21.0	18.0	Pantanello Sanctuary	6th–3rd c. BC
Cervus elaphus	phalanx proximalis	50.5	21.5	25.0	16.5	15.0	21.0	18.0	Incoronata	8th–6th c. BC
Cervus elaphus	phalanx proximalis	51.0	20.0		17.0	14.0	19.5	16.0	Pantanello Kiln Deposit	2nd c. BC–1st c. AD
Cervus elaphus	phalanx proximalis	51.0	21.5	26.0	18.0	15.0	20.0	17.0	Sant' Angelo Grieco	6th c. BC–1st c. AD
Cervus elaphus	phalanx proximalis	53.0	23.0	27.0	18.0	16.0	21.0	18.5	Pantanello Sanctuary	6th–3rd c. BC
Cervus elaphus	phalanx proximalis	53.0	24.0	28.0	18.0	15.5	22.0	18.5	Incoronata	8th–6th c. BC
Cervus elaphus	phalanx proximalis	55.0			20.0	16.3	22.5		Pantanello Sanctuary	6th–3rd c. BC
Cervus elaphus	phalanx media	36.0	21.0	27.7	15.5	19.0	19.0	26.0	Pantanello Sanctuary	6th–3rd c. BC
Cervus elaphus	phalanx media	37.0	20.8	27.0	16.0	18.0	18.0	24.0	Termitito	Late Bronze Age
Cervus elaphus	phalanx media	37.0	21.0	28.0	15.0	21.0	19.0	27.0	Termitito	Late Bronze Age
Cervus elaphus	phalanx media	38.7	21.0	29.0	15.3	20.0	19.5	26.3	Pantanello Sanctuary	6th–3rd c. BC
Cervus elaphus	phalanx media	40.0	20.5	26.0	15.0	18.0	17.5	24.0	Pantanello Kiln Deposit	2nd c. BC–1st c. AD
Cervus elaphus	phalanx media	41.0			16.0	19.0	17.7	24.0	Incoronata	8th–6th c. BC
Cervus elaphus	phalanx distalis	45.0	21.0		29.0				Incoronata	8th–6th c. BC
Cervus elaphus	phalanx distalis	46.0	21.0		28.0				Termitito	Late Bronze Age
Cervus elaphus	phalanx distalis	49.0	19.0		28.0				Pantanello Sanctuary	6th–3rd c. BC
Species	**Skeletal part**	**GL**	**Bp**	**Dp**	**SD**	**DD**	**Bd**	**Dd**	**Site name**	**Period**
Dama dama	tibia				25.0	23.0	44.0	35.0	Pantanello Kiln Deposit	2nd c. BC–1st c. AD
Dama dama	metatarsus				16.0				Pantanello Kiln Deposit	2nd c. BC–1st c. AD
Species	**Skeletal part**			**SLC**	**BG**		**GLP**		**Site name**	**Period**
Sus scrofa	scapula				25.7		39.0		Termitito	Late Bronze Age

	Skeletal part	**GL**	**Bp**	**Dp**	**SD**	**DD**	**Bd**	**Dd**	**Site name**	**Period**
Sus scrofa	humerus						42.0	43.0	San Biagio	3rd–4th c. AD
Sus scrofa	humerus						51.0		Pantanello Kiln Deposit	2nd c. BC–1st c. AD
Lepus europeus	humerus				5.0	5.0	11.5	8.5	Incoronata	8th–6th c. BC
Lepus europeus	femur						18.0	18.0	San Biagio	3rd–4th c. AD
Lepus europeus	tibia						14.0	10.0	Pantanello Pits	Late Neolithic
Species	**Skeletal part**		**P1–P4 length**	**M1–M3 length**	**M1 length**				**Site name**	**Period**
Vulpes vulpes	mandibula		29.0	24.0	13.0				Pantanello Pits	Late Neolithic
Vulpes vulpes	mandibula		29.0	24.0	14.0				Pantanello Kiln Deposit	2nd c. BC–1st c. AD
Vulpes vulpes	mandibula			23.5	14.0				Pantanello Kiln Deposit	2nd c. BC–1st c. AD
	Skeletal part			**SLC**	**BG**		**GLP**		**Site name**	**Period**
Vulpes vulpes	scapula	72.5		10.7	15.5		17.0		Pantanello Kiln Deposit	2nd c. BC–1st c. AD
	Skeletal part	**GL**	**Bp**	**Dp**	**SD**	**DD**	**Bd**	**Dd**	**Site name**	**Period**
Vulpes vulpes	humerus	108.0			73.0	8.0	19.7		Pantanello Kiln Deposit	2nd c. BC–1st c. AD
Vulpes vulpes	radius	102.0	10.8	7.0	7.0	4.0	14.0	8.5	Pantanello Kiln Deposit	2nd c. BC–1st c. AD
Vulpes vulpes	femur		23.5	11.0	8.2	7.5			Pantanello Kiln Deposit	2nd c. BC–1st c. AD
Species	**Skeletal part**	**I1–I3 length**	**P1–P4 length**	**M1–M3 length**					**Site name**	**Period**
Canis lupus	mandibula	8.5	42.0	39.0					Pantanello Sanctuary	6th–3rd c. BC
	Skeletal part			**SLC**	**BG**		**GLP**		**Site name**	**Period**
Canis lupus	scapula			21.0	29.0		36.0		Pantanello Sanctuary	6th–3rd c. BC
Canis lupus	scapula			23.0			37.0		Pantanello Necropoleis	6th– 3rd c. BC
	Skeletal part	**GL**	**Bp**	**Dp**	**SD**	**DD**	**Bd**	**Dd**	**Site name**	**Period**
Canis lupus	humerus				15.0	16.0	40.0	33.0	Pantanello Necropoleis	6th– 3rd c. BC
	Skeletal part	**GL**	**Bp**	**Dp**	**SD**	**DD**	**Bd**	**Dd**	**Site name**	**Period**
Canis lupus	radius			15.0	15.0	10.0	27.5	15.5	Pantanello Necropoleis	6th– 3rd c. BC

	Skeletal part	GL	Bp	Dp	SD	DD	Bd	Dd	Site name	Period
Canis lupus	femur		44.0	24.0	16.0	17.0	36.0	37.0	Pantanello Necropoleis	6th– 3rd c. BC
Canis lupus	tibia	200.0	41.0	42.0	15.5	15.0	25.0	21.0	Pantanello Necropoleis	6th– 3rd c. BC
	Skeletal part	**GL**	**Bd**		**Dm**				**Site name**	**Period**
Canis lupus	astragalus	34.0	23.5		18.0				Pantanello Necropoleis	6th– 3rd c. BC
	Skeletal part	**GL**	**GB**		**Greatest depth**				**Site name**	**Period**
Canis lupus	calcaneus	51.0	20.5		24.5				Pantanello Necropoleis	6th– 3rd c. BC
Species	**Skeletal part**	**GL**	**Bp**	**Dp**	**SD**	**DD**	**Bd**	**Dd**	**Site name**	**Period**
Meles meles	humerus						32.5		Pantanello Kiln Deposit	2nd c. BC– 1st c. AD
Species	**Skeletal part**			**SLC**	**BG**		**GLP**		**Site name**	**Period**
Felis catus	scapula			11.0	13.0		16.0		Termitito	Late Bronze Age

References

Albarella, U. 1997. Crane and Vulture at an Italian Bronze Age Site. *International Journal of Osteoarchaeology* 7: 346–349.

André-Salvini, B., and Descamps-Lequime, S. 2003. L'osselet de Suse: une prise de guerre antique. *Actualité du departement des Antiquités greques, étrusques et romaines* 10: 1–7.

Anderson, J. K. 1961. *Ancient Greek Horsemanship.* Berkeley and Los Angeles: University of California Press.

Azzaroli, A. 1972. II cavallo domestico in Italia dall'età del bronzo agli Etruschi. *Studi Etruschi* 40: 273–308.

———. 1980. Venetic Horses from Iron Age Burials at Padova. *Rivista di Scienza Preistoriche* 35/1–2: 281–308.

———. 1985. *An early history of horsemanship.* Leiden: E. J. Brill.

Bartosiewicz, L. 1999a. Aurochs (*Bos primigenius bojanus*, 1827) in the Holocene of Hungary. In *Archäologie und Biologie des Auerochsen,* Wissenschafliche Schriften 1, ed. G.-C. Weniger, 103–117. Mettmann: Neanderthal Museum.

———. 1999b. The Role of Sheep versus Goat in Meat Consumption at Archaeological Sites. In *Transhumant Pastoralism in Southern Europe,* Series Minor 11, eds. L. Bartosiewicz and H. J. Greenfield, 47–60. Budapest: Archaeolingua Kiadó.

———. 2005. Plain Talk: Animals, Environment and Culture in the Neolithic of the Carpathian Basin and Adjacent Areas. In *(Un)settling the Neolithic,* eds. D. Bailey and A. Whittle, 51–63. Oxford: Oxbow Books.

———. 2006. *Régenvolt háziállatok* [*Ancient Domestic Animals*]. Budapest: L'Harmattan.

Bartosiewicz, L., and Bökönyi, S. 2006. Animal Remains from Late Neolithic Causeway Camps in Western France. In *Font-Rase á Barbezieux et Font-Belle á Segonzac (Charente). Deux sites du Néolithique récent saintongeais Matignons/Peu-Richard,* British Archaeological Reports International Series 1562, ed. C. Burnez, 326–335. Oxford: Archaeopress.

Bartosiewicz, L., Van Neer, W., and Lentacker, A. 1993. Metapodial Asymmetry in Cattle. *International Journal of Osteoarchaeology* 3/2: 69–76.

———.1997. *Draught Cattle: Their Osteological Identification and History.* Annalen, Zoologische Wetenschappen Vol. 281. Tervuren: Koninklijk Museum voor Midden-Afrika.

Bartosiewicz, L., and Gál, E. 2007. Sample Size and Taxonomic Richness in Mammalian and Avian Bone Assemblages from Archaeological Sites. *Archeometriai Műhely* 2007/1:37–44; http://www.ace.hu/am2007_1/AM-2007-01-BL.pdf.

Béal, J. C. 1984. *Les objects de tabletterie antique du musée archéologique de Nimes.* Cahiers des musées et monuments de Nîmes 2.

Becker, C. 1986. *Kastanas, Ausgrabungen in einem Siedlungshügel, der Bronze- und Eisenzeit Makedoniens 1975–1979, Die Tierknochenfunde.* Prähistorische Archäologie in Südosteuropa, 5. Berlin: Volker Spiess.

———. 2005. Birds, Mice, and Slaughter Refuse from an Islamic Mosque in Syria—a Puzzling Mixture at a Peculiar Location. In *Feathers, Grit and Symbolism. Birds and Humans in the Ancient Old and New Worlds.* Proceedings of the 5th Meeting of the ICAZ Bird Working Group, Munich, Germany, 26–30 July, 2004, Documenta Archaeobiologiae 3, eds. G. Grupe and J. Peters, 271–280. Rahden/Westf.

Bejenaru, L. and Monah, D. 2006. Use of Astragals in the Rituals of the Chalcolithic Cucuteni Culture. Case Study: the Poduri-Dealul Ghundaru Tell. In *International Council for Archaeozoology 10th Conference, Mexico City 23–28 August 2006,* eds. O. J. Polaco, J. Arroyo-Cabrales, F. J. Aguilar, and A. F. Guzmán, 29. Mexico City: Instituto Nacional de Antropología e Historia.

Beldiman, C. 2002. Sur la typologie des outils en matières dures animales du Néolithique ancien de Roumanie: le poinçon sur demi-métapode perforé. In *Ateliers et techniques artisanales. Contributions archéologiques,* Bibliotheca Musei Bistriţa, Série Historica 6, ed. C. Gaiu, 14–31. Cluj-Napoca: Musée Départemental Bistriţa-Năsăud.

Benac, A., and Brodar, M. 1958. Crvena Stijena, 1956. *Glasnik Zemaljskog muzeja u Sarajevo,* XIII 33-35: 54–56.

Benda, B. 2004. *Étkezési szokások a 17. századi főúri udvarokban Magyarországon* [*Eating Customs in 17th c. Aristocratic Courts in Hungary*]. Unpublished PhD thesis. Budapest: Eötvös Loránd University.

Blanc, G. A., and Blanc, A. C. 1958–59. Il Bove della Stipe votive del Niger Lapis del Foro Romano. *Bulletino di Paletnologia Italiana* 12/67-68: 7–57.

Bökönyi, S. 1952. Les chevaux scythiques du cimetiére de Szentes–Vekerzug. *Acta Archaeologica Academiae Scientiarum Hungaricae* 2: 173–183.

———. 1954. Les chevaux scythiques de Szentes–Vekerzug. II. *Acta Archaeologica Academiae Scientiarum Hungaricae* 4: 93–114.

———. 1962a. Segasniat stadij na izsledovannijata verhu istoriata na domasnite zhivotni [On the Current State of Research into Domestication and the History of Domestic Animals]. *Priroda* 11: 7–13.

———. 1962b. Zur Naturgeschichte des Ures in Ungarn und das Problem der Domestikation des Hausrindes. *Acta Archaeologica Academiae Scientiarum Hungaricae 14*: 175–214.

———. 1964. Angaben zur Kenntnis der eisenzeitlichen Pferde in Mittel- und Osteuropa. *Acta Archaeologica Academiae Scientiarum Hungaricae* 16: 227–239.

———. 1965. Vergleichende Untersuchung der Pferdeskelette des baierischen Gräberfeldes von Linz-Zizlau I. *Naturkundliche Jahrbuch der Stadt Linz* 1965: 7–20.

———. 1968. *Data on Iron Age Horses of Central and Eastern Europe.* American School of Prehistoric Research, Peabody Museum, Harvard University, Bull. 25. Cambridge, Mass.1–71.

———. 1970. Animal remains from Lepenski Vir. *Science* 167 (3263): 1702–1704.

———. 1972. Once More on the Osteological Differences of the Horse, the Half-ass and the Ass. In *The Caspian Miniature Horse in Iran*, Field Research Projects 64, ed. L. Firouz, 12–23. Miami: University of Miami Press.

———. 1974. *History of domestic mammals in Central and Eastern Europe.* Budapest: Akadémiai Kiadó.

———. 1976. The Vertebrate Fauna of Obre. *Wissenschaftliche Mitteilungen des bosnisch-herzegowinischen Landesmuseums* 4/A: 55–154.

———. 1977. *Animal Remains from the Kermanshah Valley, Iran.* British Archaeological Reports, Suppl. Series 34.

———. 1977–82. The Early Neolithic Fauna of Rendina. *Origini* XI. Roma: 345–354.

———. 1978a. The Introduction of Sheep Breeding to Europe. *Ethnozootechn* 21: 65–70.

———. 1978b. The Vertebrate Fauna of Vlasac. In *Vlasac; a Mesolithic Settlement in the Iron Gate,* Serbian Academy of Sciences and Arts, Monograph DXII, ed. D. Srejović and Z. Letica, 35–65. Beograd.

———. 1981. Mende–Leányvár Árpád-kori–13. Századi–állatmaradványai (Árpádenzeitliche 13. Jh. Tierreste aus Mende–Leányvár). *Archeológiai Értesítő* 108/2: 251–258.

———. 1983a. Animal Bones from Test Excavations of Early Neolithic Ditched Villages on the Tavoliére, South Italy. In *Studi sul neolitico del Tavoliére della Puglia. Indagine territoriale in un'area campione,* British Archaeological Reports, International Series 160, eds. S. M. Cassano and A. Manfredini, 237–249. Oxford: Archaeopress.

———. 1983b. Late Chalcolithic and Early Bronze Age I Animal Remains from Arslantepe (Malatya): A Preliminary Report. *Origini* XII. 2: 581–597.

———. 1983c. Domestication, Dispersal and Use of Animals in Europe. In *World Animal Science, Subseries* A: *Basic Information* 1, eds. L. Peel and D. E. Tribe, 1–20. Amsterdam: Elsevier.

———. 1984. *Animal Husbandry and Hunting in Tác–Gorsium, The Vertebrate Fauna of a Roman Town in Pannonia.* Studia Archaeologica VIII. Budapest: Akadémiai Kiadó.

———. 1985. A Comparison of the Early Neolithic Domestic and Wild Faunas of the Balkans, Italy and South France. *Cahiers Ligures de Préhistoire et de Protohistoire.* N. S. 2: 181–192.

———. 1986a. The Equids of Umm Dabaghiyah, Iraq. In *Equids in the Ancient World.* Beihefte zum Tübinger Atlas des Vorderen Orients, A. Nr. 19/1, eds. R. H. Meadow and H.-P. Uerpmann, 302–317. Wiesbaden: Dr. Ludwig Reichert Verlag.

———. 1986b. Animal Remains from the Roman Villa of San Potito-Ovindoli (L'Aquila) 1983–84. (A Preliminary Report.) *Acta Archaeologica Academiae Scientiarum Hungaricae* 38: 89–91.

———. 1986c. Faunal Remains. In *Excavations at Sitagroi. A Prehistoric Village in Northeast Greece.* Volume 1, eds. C. Renfrew, M. Gimbutas and E. S. Elster, 63–132. Los Angeles: The University of California.

———. 1987. Horses and Sheep in East Europe in the Copper and Bronze Ages. In *Proto-Indo-European: The Archaeology of a Linguistic Problem. Studies in Honor of Maria Gimbutas,* eds. S. N. Skomal and E. G. Polomé, 136–144. Washington, D. C.: Institute for the Study of Man.

———. 1990. The Fauna. In *Roccagloriosa I. L'abitato: scavo e ricognizione topografica (1976–1986),* ed. G. Tocco Sciarelli, 329–331. Naples: Publications du Centre Jean Bérard.

———. 1992. Animal Remains from Mihajlovac–Knjepište: an Early Neolithic Settlement of the Iron Gate Gorge. *Balcanica* XXIII: 77–87.

———. 1998. Faunal Remains (Appendix 11A. 6). In *The Chora of Metaponto: The Necropoleis* Vol. II, ed. J. C. Carter, 560–562. Austin: University of Texas Press.

Bökönyi, S., Costantini, L., and Fitt, J. 1993. The Farming Economy. In *Fourth Century BC Magna Grecia: A Case Study,* ed. M. Gualtieri, 281–290. Jonsered: Paul Åströms Förlag.

Bökönyi, S., and Kubasiewicz, M. 1961. *Neolithische Tiere Polens und Ungarns in Ausgrabungen. I. Das Hausrind.* Budapest–Szczecin: Panstwowe Wydawnictwo Naukowe.

Bökönyi, S., and Siracusano, G. 1987. Reperti faunistici dell' età del bronzo del sito di Coppa Nevigata: un commento preliminare. In *Coppa Nevigata il suo territorio. Testimonianze archeologiche dal VII al II millennio a. C.*, eds. S. M. Cassano, A. Cazella, A. Manfredini and M. Moscoloni, 205–210. Roma: Quasar.

Bollongino, R., Edwards, C. J., Alt, K. W., Burger, J., and Bradley, D. G. 2006. Early History of European Domestic Cattle as Revealed by Ancient DNA. *Biology Letters* 2: 155–159.

Braidwood, R. J., and Reed, C. A. 1957. The Achievement and Early Consequences of Food Production: a Consideration of the Natural-Historical Evidence. *Cold Springs Harbor Symposia on Quantitative Biology* 22: 19–31.

Brauner, A. A. 1916. *Materiali k poznaniyu domasnikh zhivotnikh Rossii. 1. Loshad kurgannikh pogrebenij Tiraspolskogo uezda Hersonskoj gubernii, Equus goskewitsch, mihi.* [Material on the Knowledge of Domestic Animals in Russia. 1. A Horse in a Kurgan Burial in the Region of Tiraspol' in the Province of Hersonskoj.] Zapiski Imperialnogo Obshchestva Selskogo Hozyaistva Juzhnoy Rossii 86, 1–152. Odessa.

Brentjes, B. 1965. Die Haustierwendung in Orient. Ein archäologischer Beitrag zur Zoologie. Wittenberg: Die Neue Brehm-Bücherei, 344.

van den Brink, F. H. 1957. *Die Säugetiere Europas.* Hamburg, Berlin: Paul Parey.

Cabaniss, B. 1983. Preliminary Faunal Report, Kiln Deposit, Pizzica-Pantanello, 1981. In *The Territory of Metaponto 1981–1982*, ed. J. C. Carter, 43–45. Austin. Available through the David Brown Publishing Company and at utexas.edu/research/ica/publications/ICA_publ.htm.

Camps-Fabre, H., and D'Anna, A. 1977. Fabrication expérimentale d'outils partir métapodes de mouton et tibias de lapin. In *Méthodologie appliquée a l'industrie de l'os préhistorique*, ed. H. Camps-Fabre, 311–323. Paris: Colloques Internationaux du C. N. R. S., 568.

Carter, J. C. 1983a. Pizzica-Pantanello. The Deposit of Pottery and Bone, 1980–81. In *The Territory of Metaponto 1981–1982*, 39–41. Austin. Available through the David Brown Publishing Company and at utexas.edu/research/ica/publications/ICA_publ.htm.

———.1983b. Necropolis at Pizzica-Pantanello. In *The Territory of Metaponto 1981–1982*, 39–41. Austin. Available through the David Brown Publishing Company and at utexas.edu/research/ica/publications/ICA_publ.htm.

———.1998a. *The Chora of Metaponto: The Necropoleis*, 2 vols. Austin: University of Texas Press.

———. 1998b. Horse Burial and Horsemanship in Magna Grecia. In *Man and the Animal World. Studies in Memoriam Sándor Bökönyi*, Main Series 8, ed. P. Anreiter, L. Bartosiewicz, E. Jerem, and W. Meid, 131–146. Budapest: Archaeolingua Kiadó.

———.1990. Metapontum: Land, Wealth, and Population. In *Greek Colonists and Native Populations*, Proceedings of the First Australian Congress of Classical Archaeology held in Honour of Emeritus Professor A. D. Trendall, Sydney 9–14 July 1989, ed. J-P. Descoeudres, 405–441. Oxford: Clarendon Press.

———. 2006. *Discovering the Greek Countryside at Metaponto.* Thomas Spencer Jerome Lectures, 23rd Series. Ann Arbor: University of Michigan Press.

Cassoli, P. F., and Tagliacozzo, A. 1997. Butchering and Cooking of Birds in the Paleolithic Site of Grotta Romanelli. *International Journal of Osteoarchaeology* 7: 303–320.

Choyke, A. M. 1996. Worked Animal Bone at the Sarmatian Site Gyoma 133. In *Cultural and Landscape Changes in South-East Hungary*, Main Series 5, ed. S. Bökönyi, 307–322. Budapest: Archaeolingua.

———. 1997. The Manufacturing Continuum. *Anthropozoologica* 25–26: 65–72.

Costantini, L. 2003. Agriculture and Diet in the Chora of Metaponto: The Paleobotanical Evidence from Pantanello. In *Living off the Chora. Diet and Nutrition at Metaponto*, ed. J. C. Carter, 3–12. Institute of Classical Archaeology, the University of Texas at Austin

Craig, O. E., Chapman, J., Heron, C., Willis, L. H., Bartosiewicz, L., Taylor, G., Whittle, A., and Collins, M. 2005. Did the First Farmers of Central and Eastern Europe Produce Dairy Foods? *Antiquity* 79: 882–894.

Cramp, S. ed. 1998. *The Complete Birds of the Western Palearctic on CD-ROM.* Oxford: Oxford University Press.

Crosby, A. W. 1986. *Ecological Imperialism: The Biological Expansion of Europe 900–1900 AD.* Cambridge: Cambridge University Press.

Cruz-Uribe, K. 1990. Animal Bones. In *La Muculufa, the Early Bronze Age Sanctuary: The Early Bronze Age Village. Excavations of 1982 and 1983*, eds. R. Ross Holloway, M. S. Joukowsky, and S. S. Lukesh, 51–64. Providence, Rhode Island: Centre for Old World Archaeology and Art, Brown University.

———. 1995. Appendix II: Preliminary Report on the Faunal Remains. In *Ustica I*, Archaeologia Transatlantica XIV, eds. R. R. Holloway and S. S. Lukesh, 91–97. Louvain-la-Neuve: Publications d'histoire de l'Art et d'Archéologie de l'Université Catholique de Louvain.

Csányi, M., and Tárnoki, J. 1992. Katalog der ausgestellten Funde. In *Bronzezeit in Ungarn. Forschungen in Tell-Siedlungen an Donau und Theiss*, eds. W. Meier-Arendt, 175–210. Frankfurt am Main: Museum für Vor- und Frühgeschichte-Archäologisches Museum–Pytheas GmbH.

D'Annibale, C. 1983. Field Survey of the Chora of Metaponto, 1981–82. In *The Territory of Metaponto 1981–82*, ed.

J. C. Carter, 7–11. Austin. Available through the David Brown Publishing Company and at utexas.edu/research/ica/publications/ICA_publ.htm.

Deschler-Erb, S. *Römische Beinartefakte aus Augusta Raurica,* Forschungen in Augst 27. Augst: Römermuseum Augst.

Di Rosa, M. 2003. I resti faunistici. In *Monte Maranfusa. Un insediamento nella Media Valle del Belice,* ed. F. Spatafora, 397–413. Palermo: Assessorato Regionale dei Beni Culturali Ambientali e della Pubblica Istruzione.

Douglas, N. 1928. *The Birds and Beasts of the Greek Anthology.* London: Chapman and Hall Ltd.

Driesch, A. von den. 1975. Die Bewertung pathologischanatomischer Veränderungen an vor- und frühgeschichtlichen Tierknochen. In *Archaeozoological Studies,* ed. A. T. Clason, 413–425. Amsterdam: North Holland Publishing Company.

———. 1976. *A Guide to the Measurement of Animal Bones from Archaeological Sites,* Peabody Museum Bulletin 1. Cambridge, MA: Harvard University.

Driesch, A. von den, and Boessneck, J. 1983. A Roman Cat Skeleton from Quseir on the Red Sea Coast. *Journal of Archaeological Science* 10: 205–211.

Erickson, B. L. 1998. Glass, Bone Artifacts, and Jewelry in Terracotta. In *The Chora of Metaponto: The Necropoleis,* Vol II, ed. J. C. Carter, 834–840. Austin: University of Texas Press.

Esteban Nadal, M., and Carbonell Roure, E. 2004. Sawtoothed Sickles and Bone Anvils: A Medieval Technique from Spain. *Antiquity* 78: 637–646.

Fernández, H., and Monchot, H. 2007. Sexual Dimorphism in Limb Bones of Ibex (*Capra ibex* L.): Mixture Analysis Applied to Modern and Fossil Data. *International Journal of Osteoarchaeology* 17: 479–491.

Forchhammer, G., Steenstrup, J., and Worsaae, J. 1851–1856. *Undersøgelser i geologisk-antikvarisk retning.* København: Kongliga Hofbogtrykker Bianco Luno.

Gaffrey, G. 1961. *Merkmale der wildlebenden Säugetiere Mitteleuropas.* Leipzig: Akademische Verlagsgesellschaft.

Gál, E. 2005. New Evidence of Fowling and Poultry Keeping in Pannonia, Dacia and Moesia during the Period of the Roman Empire. In *Feathers, Grit and Symbolism. Birds and Humans in the Ancient Old and New Worlds,* Proceedings of the 5th Meeting of the ICAZ Bird Working Group, Munich, Germany, 26–30 July, 2004, Documenta Archaeobiologiae 3, ed. G. Grupe and J. Peters, 303–318. Rahden/Westf.: Verlag Marie Leidorf GmbH.

———. 2007. *Fowling in Lowlands. Neolithic and Copper Age Bird Bone Remains from the Great Hungarian Plain and South-East Romania.* Series Minor 24, Budapest: Archaeolingua.

———. 2008a. Nuove ossa animali dalla villa romana (I–III sec. d. C.) di San Potito d'Ovindoli (L'Aquila, Italia). In *Richerche archeologiche a San Potito di Ovindoli e nell'aree limitrofe,* eds. D. Gabler and F. Redő, 179–193, L'Aquila: Ricerche Etno Antropologiche.

———. 2008b (forthcoming). Faunal Analysis: Bones from Mammals of Economic Importance. In J. Morter, *The Chora of Croton 1: The Neolithic Settlement at Capo Alfiere,* ed. J. Robb.

———. 2008c. Faunal and Taphonomic Approach to a Late Pleistocene Cavern Bird Bone Assemblage in North-west Hungary. *Geobios* 41: 79–90.

———. (2008d, in press). Broken-winged: Fossil and Subfossil Pathological Bird Bones from Recent Excavations. In *Proceedings of the 1st Meeting of the ICAZ Palaeopathology Working Group, Nitra 23–24 September 2004,* British Archaeological Reports, International Series, eds. R. Thomas and Z. Miklíková. Oxford: Archaeopress.

———. 2008e. Bone Evidence of Pathological Lesions in Domestic Hen (*Gallus domesticus* Linnaeus, 1758). *Veterinarija ir Zootechnika* 41(63): 42–48.

———. (2008f, forthcoming). Bone Artifacts. In J. Morter, *The Chora of Croton 1: The Neolithic Settlement at Capo Alfiere,* ed. J. Robb.

Gandert, O.-F. 1952. Zur Abstammungs- und Kulturgeschichte des Haushuhns. *Actes du IVe Congrès international des sciences anthropologiques et ethnologiques* II: 113–118.

———. 1953. Zur Abstammungs- und Kulturgeschichte des Hausgeflügels, insbesonders des Haushuhnes. In *Beiträge zur Frühgeschichte der Landwirtschaft,* I, eds. W. Rothmaler and W. Padberg, 69–81. Berlin: Deutsche Akademie der Landwirtschaftswissenschaften.

Garcia Petit, L. 2005. Recent Studies on Prehistoric to Medieval Bird Bone Remains from Catalonia and Southeast France. In *Feathers, Grit and Symbolism. Birds and Humans in the Ancient Old and New Worlds.* Proceedings of the 5th Meeting of the ICAZ Bird Working Group, Munich, Germany, 26–30 July, 2004, Documenta Archaeobiologiae 3, eds. G. Grupe and J. Peters, 147–163. Rahden/Westf: Verlag Marie Leidorf GmbH.

Geppert, P. 1990. *Hundeschlachtungen in Deutschland im 19. und 20. Jahrhundert unter besonderer Berücksichtigung der Verhältnisse in München.* München: Institut für Paläoanatomie, Domestikationsforschung und Geschichte der Tiermedizin der Universität München.

Giomi, F. 1996. Favella, industria in osso. In *Forme e tempi della neolitizzazione in Italia meridionale e in Sicilia,* ed. V. Tiné, 361–363. Atti del seminario internazionale de Rossano, 29 aprile – 2 maggio 1994, Tomo I, Istituto Regionale per le Antichitá Calabresi.

Green, M. W. 1980. Animal Husbandry at Uruk in the Archaic Period. *Journal of Near Eastern Studies* 39: 1–19.

Green, P. 1991. *Alexander of Macedon, 356–323 BC: A Historical Biography.* Berkeley: University of California Press.

Grüss, J. 1933. Über Milchreste aus der Hallstattzeit und andere Funde. *Forschungen und Fortschritte* 9: 105–106.

Gulde, V. 1985. *Osteologische Untersuchungen an Tierknochen aus dem römischen Vicus von Rainau-Buch (Ostalbkreis).* Stuttgart: Materialhefte zur Vor- und Frühgeschichte in Baden-Württemberg 5.

Habermehl, K. H. 1957. Die Tierknochenfunde im römischen Lagerdorf Butzbach. *Saalburg-Jahrbuch* 16: 67–108.

Hančar, Fr. 1955–56. *Das Pferd in prähistorischer und früher historischer Zeit.* Wien–München: Wiener Beiträge zur Kulturgeschichte und Linguistik, Bd. 11.

Harcourt, R. A. 1974. The Dog in Prehistoric and Early Historic Britain. *Journal of Archaeological Science* 1: 151–175.

Harmatta, J. 1968. Früheisenzeitliche Beziehungen zwischen dem Karpatenbecken, Oberitalien und Griechenland. *Acta Archaeologica Academiae Scientiarum Hungaricae* 20: 153–157.

Hemmer, J., and Eichmann, M. 1977. Hunde aus der Römerzeit des Rhein-Main-Gebietes. *Mainzer Naturwissenschaftliches Archiv* 11: 257–274.

Hilzheimer, M. 1932. Römische Hundeschädel aus Mainz, ein fränkischer Hundeschädel und ein Hundeschädel des 15. oder 16. Jahrhunderts ebendaher. *Biologia generalis* 8: 91–126.

Hole, F., Flannery, K. V. 1967. The Prehistory of Southwestern Iran: A Preliminary Report. *Proceedings of the Prehistoric Society* XXXIII: 147–206.

Holloway, R. R. 1975. Buccino: the Early Bronze Age village of Tufariello. *Journal of Field Archaeology* 2: 11–81.

Hornberger, M. 1970. *Gesamtbeurteilung der Tierknochenfunde aus der Stadt auf dem Magdalensberg in Kärnten (1948–1966).* Klagenfurt: Kärntner Museumsschrifte XLIX.

Jameson, M. H. 1988. Sacrifice and Animal Husbandry in Classical Greece. In *Pastoral Economies in Classical Antiquity,* Cambridge Philological Society, Supplementary Volume no. 14, ed. C. R. Whittaker, 87–119. Cambridge.

Jarman, M. R. 1972. European Deer Economies and the Advent of the Neolithic. In *Papers in Economic Prehistory,* ed. E. S. Higgs, 125–147. Cambridge: Cambridge University Press.

Jourdan, L. 1976. *La faune du site gallo-romain et paléochrétien de la Bourse (Marseille).* Dissertation, Paris.

Kardos, L. 1943. *Az Őrség népi táplálkozása* [*Traditional Food in the Őrség Region*]. Budapest: Államtudományi Intézet Táj- és Népkutató Osztálya.

Keller, O. 1909–1913. *Die Antike Tierwelt,* 2 vols. Leipzig: Wilhelm Engelmann.

Kokabi, M. 1982. *Arae Flaviae II, Viehhaltung und Jagd im römischen Rottweil.* Stuttgart: Forschungen und Berichte Zur Vor- und Frühgeschichte in Baden Württemberg, 13.

Koudelka, F., 1885. Das Verhältnis der Ossa longa zur Skeletthöhe bei den Säugetieren. *Verhandlungen des naturforschenden Vereins Brünn* 24: 127–153.

Kovács, Gy. 1989. Juh astragalus-játékkockák a szolnoki vár területéről [Astragali aus dem Gebiet der Burg von Szolnok]. *Archeológiai Értesítő* 114/1–2: 103–110.

Kratochvíl, Z. 1976. Das postkraniale Skelett der Wild- und Hauskatze (*Felis silvestris und F. Lybica f. Catus*). *Acta Scientiarum Naturalium Academiae Scientiarum Bohemoslovacae Nova Series* X. 6: 1–43.

La Baume, W. 1947. Diluviale Schädel vom Ur (*Bos primigenius Bojanus*) aus Toscana. Berichte der Schweizerischen Paläontologischen Gesesellschaft 26. Jahresversammlung. *Eclogae geologicae Helvetiae* 40: 299–308.

Lehman, U. 1949. Der Ur im Diluvium Deutschlands und seine Verbreitung. *Neues Jahrbuch für Mineralogie, Geologie und Paläontologie* 90 B: 163–266.

Leithner, O. von, 1927. Der Ur. *Berichte der Internationalen Gesellschaft zur Erhaltung des Wisents* 2: 1–139.

Lengerken, H. 1955. *Ur, Hausrind und Mensch.* Wissenschaftliche Abhandlungen Nr. 14. Berlin: Deutsche Akademie der Landwirtschaftswissenschaften.

Lüttschwager, J. 1966. Über ein Hundeskelett aus einer Römersiedlung in Heilbronn am Neckar. *Säugetierkundliche Mitteilungen* 14: 85–91.

Mann, C. C. 2007. The Legacy of Jamestown: America Found and Lost. *National Geographic,* May 2007: 32–55.

Marković, C. 1974. The Stratigraphy and Chronology of the Odmut Cave. *Archaeologia Yugoslavica* XV: 7–12.

Matolcsi, J. 1970. Historische Erforschung der Körpergrösse des Rindes auf Grund von ungarischen Knochenmaterial. *Zeitschrift für Tierzüchtung und Züchtungsbiologie* 87/2: 89–137.

Moreno-García, M., Pimenta, C., and Gros, M. 2005. Musical Vultures in the Iberian Peninsula: Sounds through their Wings. In *Feathers, Grit and Symbolism. Birds and Humans in the Ancient Old and New Worlds.* Proceedings of the 5th Meeting of the ICAZ Bird Working Group, Munich, Germany, 26–30 July, 2004, Documenta Archaeobiologiae 3, eds. G. Grupe and J. Peters, 329–347. Rahden/Westf.: Verlag Marie Leidorf GmbH.

Moreno-García, M., Pimenta, C. M., and Ruas, J. P. 2005. Safras em osso para picar foicinhas de gume serrilhado… a sua longa história! *Revista Portuguesa de Arqueologia* 8(2): 571–627.

Moreno-García, M., Pimenta, C. M., López Aldana, P. M. and Pajuelo Pando, A. 2007. The Signature of a Blacksmith on a Dromedary Bone from Islamic Seville (Spain). *Archaeofauna* 16: 193–202.

Morrison, E. 2007. *Beasts. Factual and fantastic.* Los Angeles: Paul Getty Museum.

Muray, C. 1979. Les techniques de débitage de métapodes de petits ruminants a Auvernier-Port. In *L'industrie en os et bois de Cervidé durant le Néolitique et l'âge des Métaux,* ed. H. Camps-Fabrer, 27–35. Paris: Éditions du Centre National de la Recherche Scientifique.

Nobis, G. 1954. Zur Kenntnis der ur- und frühgeschichtlichen Rinder Nord- und Mitteldeutschlands. *Zeitschrift für Tierzüchtung und Züchtungsbiologie* 63: 155–194.

Payne, S. 1973. Animal bones (of the Franchthi Cave). *Hesperia* XLII (1): 59–66.

Peterson, R. T., Mountfort, G., and Hollom, P. A. D. 1966. *A Field Guide to the Birds of Britain and Europe.* London: Collins.

Pétrequin, P., Arbogast, R-M., Péterquin, A-M., Van Willigen, S., and Bailly, M., eds. 2006. *Premiers chariots, premiers araires. La diffusion de la traction animale en Europe pendant les IVe et IIIe millénaires avant notre ère,* CRA Monographies 29. Paris: Éditions du Centre National de la Recherche Scientifique.

Pohlig, H. 1912. *Bovidés fossils de l'Italie.* Bulletin de la Société Belge de Géologie 25. Bruxelles.

Portis, A. 1907. *De alcuni avanci fossili di grandi Ruminantia principalmente della provincia di Roma.* Paleontografia Italica 13. Roma.

Proháska, M. 1998. Metal Objects and Coins. In *The Chora of Metaponto: The Necropoleis,* Vol. II, ed. J. C. Carter, 820–821. Austin: University of Texas Press.

Requate, H. 1957. Zur Naturgeschichte des Ures (Bos primigenius Bojanus 1827). Nach Schädel- und Skelettfunden in Schleswig-Holstein. *Zeitschrift für Tierzüchtung und Züchtungsbiologie* 70/4: 297–338.

Riedel, A. 1974. I mammiferi domestici della grotta N. 1745/4558 V. G. e di faune oloceniche minori. *Atti e Memorie della Commissione Grotte "Eugenio Boegan"* 13: 53–90.

———. 1976. La fauna del villaggio preistorico di Ledro. Archeo-zoologia e paleo-economia. *Studi Trentini di Scienze Naturali* 53/5B *Nuova Serie*: 3–120.

———. 1977. The fauna of four prehistoric settlements in Northern Italy. *Atti del Museo Civico di Storia Naturale di Trieste* 30/1/6: 65–122.

———. 1979. La fauna di alcuni insediamenti preistorici del territorio veronese. *Atti del Museo Civico di Storia Naturale di Trieste* 31/1/4: 41–73.

———. 1981. La fauna di Braida Roggia a Pozzuolo del Friuli. *Atti dei Civici Musei di Storia ed Arte di Trieste* 12/1: 121–131.

———. 1984. The Paleovenetian Horse of Le Brustolade (Altino). *Studi Etruschi* L: 227–256.

———. 1986. Archäozoologische Untersuchungen im Raum zwischen Adriaküste und Alpenhauptkamm (Spätneolithikum bis zum Mittelalter). *Padusa* XXII: 1–220.

Rüeger, J. 1944. Knochenfunde. In W. Drack, Das römische Ökonomiegebäude in Kirchdorf. *Argovia* 56: 236–237.

Rütimeyer, L. 1861. *Die Fauna der Pfahlbauten der Schweiz.* Neue Denkschriften der Allgemeinen Schweizerischen Gesellschaft für die Gesammten Naturwissenschaften 19. Bern: Schweizerische Akademie der Naturwissenschaften.

Scali, S. 1983. Observations on the Faunal Remains from the Territory of Metaponto. In *The Territory of Metaponto 1981–1982,* ed. J. C. Carter, 46–49. Austin. Available through the David Brown Publishing Company and at utexas.edu/research/ica/publications/ICA_publ.htm.

Schibler, J. 1980. *Osteologische Untersuchungen der cortaillodzeitlichen Knochenartefakte.* Die neolithischen Ufersiedlungen von Twann. Band 8. Bern: Staatlicher Lehrmittelverlag.

———. 1981. *Typologische Untersuchungen der cortaillodzeitlichen Knochenartefakte. Die neolithischen Ufersiedlungen von Twann.* Band 17. Bern: Staatlicher Lehrmittelverlag.

Schmid, E. 1970. Über Knochenfunde aus der römischen Stadt Augusta Raurica. *Actes du VII Congrès International des Sciences Préhistoriques et Protohistoriques, Prague 1966,* Praha 1970: 1316–1320.

Schlumbaum, A., Stopp, B., Breuer G., Rehazek, A., Turgay, M., Blatter, R., and Schibler, J. 2003. Combining Archaeozoology and Molecular Genetics: The Reason Behind the Changes in Cattle Size Between 150 BC and 700 AD in Switzerland. *Antiquity* 77: http://antiquity.ac.uk/ProjGall/schlumbaum/index.html

Schlumbaum, A., Turgay, M., and Schibler, J. 2006. Near East mtDNA Haplotype Variants in Roman Cattle from Augusta Raurica, Switzerland, and in the Swiss Evoléne Breed. *Animal Genetics* 37/4: 373–375.

Schramm, Z. 1967. Kości długie a wysokość w kłębie u kozy [Long Bones and Height in Withers in Goats]. *Roczniki Wyźsej Szkoly Rolniczej w Poznaniu* XXXVI: 89–105.

Semenov, S. A. 1964. *Prehistoric Technology.* London: Cory Adams and Nackay.

Serjeantson, D. 1997. Subsistence and Symbol: The Interpretation of Bird Remains in Archaeology. *International Journal of Osteoarchaeology* 7: 255–259.

Sidéra, I. 2005. Technical Data, Typological Data: A Comparison. In *From Hooves to Horns, from Mollusc to Mammoth. Manufacture and Use of Bone Artefacts from*

Prehistoric Times to the Present, Muinasaja teadus 15, eds. H. Luik, A. M. Choyke, C. E. Batey, and L. Lõugas, 81–90. Tallinn: Archaeological Department of the Institute of History–Chair of Archaeology of the University of Tartu.

Sorrentino, C. 1983. La fauna. In *Passo di Corvo e la civiltà neolitica del Tavolière,* ed. S. Tinè, 149–158. Genova: Sagep.

Srejović, D. 1974. The Odmut Cave—A New Fact of the Mesolithic Culture of the Balkan Peninsula. *Archaeologia Yugoslavica* XV: 3–6.

Stampfli, H. R. 1959–60. Die Tierwelt der kelto-römischen Siedlung "Engelhalbinsel" bei Bern. *Jahrbuch des Bernischen Historischen Museums in Bern* XXXIX. und XL: 415–434.

Stehlin, H. G., Graziosi, O. 1935. Ricerche sugli asinidi fossili d'Europa. *Mémoires de la Société de Paléontologie de Suisse* 56: 1–68.

Suter, P. J. 1981. Die Hirschgeweihartefakte der Cortaillod-Schichten. *Die neolitischen Ufersiedlungen von Twann.* Band 15. Bern: Staatlicher Lehrmittelverlag.

Szalay, B. 1915. *Der Wisent in Ortsnamen.* Zoologische Annalen. Würzburg: A. Stuber's Verlag.

Tagliacozzo, A. 1994a. Economic Changes between the Mesolithic and the Neolithic in the Grotta dell'Uzzo (Sicily, Italy). *The Journal of the Accordia Research Centre* 5: 7–37.

———. 1994b. I dati archeozoologici: economia di allevamento e caccia a Broglio di Trebisacce. In *Enotri e Micenei nella Sibaritide,* eds. R. Peroni and F. Trucco, 587–652. Taranto: Istituto per la storia e l'archeologia della Magna Grecia.

———. 2005–2006. Animal exploitation in the Early Neolithic in Central-Southern Italy, *Munibe (Antropologia-Arkeologia).* 57: 429–439.

Tagliacozzo, A., and Gala, M. 2002. Exploitation of Anseriformes at Two Upper Palaeolithic Sites in Southern Italy: Grotta Romanelli (Apulia) and Grotta del Santuario della Madonna a Praia a Mare (Calabria). In *Proceedings of the 4th Meeting of the ICAZ Bird Working Group, Kraków, Poland, 11–15 September, 2001,* Acta zoologica cracoviensia 45, eds. Z. M. Bocheński, Z. Bocheński and J. R. Stewart, 117–131. Kraków: Instytut Systematyki i Ewolucji Zwierząt PAN.

Takács, I. 1991. The History of Pig (*Sus scrofa dom.* L.) Butchering and the Evidence for Singeing on Subfossil Teeth. *A Magyar Mezőgazdasági Múzeum Közleményei 1990–1991:* 17–41.

Tamás, L. ed. 1987 *Állatorvosi sebészet* [*Veterinary Surgery*] 2. Budapest: Mezőgazdasági Kiadó.

Teichert, M. 1975. Osteologische Unterschungen zur Berechnung der Widerristhöhe bei Schafen. In *Archeological studies,* ed. A. T. Clason, 51–69. Amsterdam: North Holland Publishing Company.

Thompson, D'Arcy, 1895. *A Glossary of Greek Birds.* Oxford: Clarendon Press.

Tomek, T., and Bocheński, Z. M. 2000. *The Comparative Osteology of European Corvids (Aves: Corvidae), with a Key to the Identification of their Skeletal Elements.* Kraków: Wydawnictwa Instytutu Systematyki i Ewolucji Zwierząt PAN.

Toynbee, A. 1965. *Hannibal's Legacy: The Hannibalic War's Effects on Roman Life.* Oxford: Oxford University Press.

Toynbee, J. M. C. 1973. *Animals in Roman Life and Art.* London: Thames and Hudson.

Uerpmann, H.-P. 1973. Animal Bone Finds and Economic Archaeology: A Critical Study of "Osteo-archaeological" Method. *World Archaeology* 4/3: 307–322.

Vitt, V. O. 1952. Loshady pazyrykskikh kurganov. *Sovietskaya Archeologiya* 16: 163–205.

Vörös, I. 1996. Dog as Building Offering from the Bronze Age Tell at Jászdózsa. *Folia Archaeologica* 45: 69–90.

Wäsle, R. 1976. *Gebissanomalien und pathologisch-anatomische Veränderungen an Knochenfundeen aus archäologischen Ausgrabungen.* Dissertation. München: Institut für Paläoanatomie, Domestikationsforschung und Geschichte der Tiermedizin der Universität München.

West, B., Ben-Xiong Zhou, 1988. Did Chickens Go North? New Evidence for Domestication. *Journal of Archaeological Science* 15: 515–533.

Whittle, A. ed. 2007. *The Early Neolithic on the Great Hungarian Plain: Investigations of the Körös Culture Site of Ecsegfalva 23, County Békés,* Varia Archaeologica Hungarica 21. Budapest: Archaeological Institute of the Hungarian Academy of Sciences.

Wilkens, B. 1989–1990. La Fauna dei livelli neolitici della Grotta Continenza. *Rivista di Scienze Preistoriche* 42: 93–100.

Winiger, J. 1999. *Rohstoff, Form und Funktion. Fünf Studien zum Neolithikum Mitteleuropas.* British Archaeological Reports, International Series 771. Oxford: Archaeopress.

Würgler, F. E. 1959. Die Knochenfunde aus dem spätrömischen Kastell Schaan. *Jahrbuch des Historischen Vereins für das Fürstentum Lichtenstein* 58: 255–282.

Zalai-Gaál, I., and Gál, E. 2005. Gerät oder Machtabzeichen? Die Hirschgeweihäxte des transdanubischen Spätneolithikums. *Acta Archaeologica Academiae Scientiarum Hungaricae* 56: 29–66.

Zeuner, F. E. 1963. *A History of Domesticated Animals.* New York: Harper and Row.

Sándor Bökönyi
Publications Relevant to Italian Archaeology

See Bartosiewicz 1998 for a full list of publications by Sándor Bökönyi as of 1998.

Bökönyi, S. 1977–1982. The Early Neolithic Fauna of Rendina. *Origini,* XI: 345–354.

———. 1983. Animal Bones from Test Excavations of Early Neolithic Ditched Villages on the Tavoliére, South Italy. In *Studi sul neolitico del Tavoliére della Puglia. Indagine territoriale in un'area campione,* eds. S. M. Cassano and A. Manfredini. Oxford, BAR, International Series 160: 237–249.

———. 1985. The Animal Remains of Maadi, Egypt: a Preliminary Report. In *Studi di paletnologia in onore di Salvatore M. Puglisi,* eds. M. Liverani, A. Palmieri, and R. Peroni. Roma: 495–499.

———. 1985. A Comparison of the Early Neolithic Domestic and Wild Faunas of the Balkans, Italy and South France. *Cah. Ligur. de Préhist. et Protohist.,* N.S. 2: 181–192.

———. 1986. Animal Remains from the Roman Villa of San Potito-Ovindoli (L'Aquila) 1983–1984 (A Preliminary Report). *Acta Arch. Hung.,* 38: 89–91.

———. 1987. Lo sviluppo degli animali domestici come si riscontra nei resti di fauna del Castelar di Rover (Possagno-Treviso). In *La ricostruzione dell'ambiente antico attraverso lo studio e l'analisi del terreneo e dei manufatti (strumenti e metodi di ricerca),* Univ. di Padova, Semin. d. Arch. d. Venezie e Topogr. dell'Italia Antica, V, eds. P. Basso, G. Bertoldo, and I. Riera. Padova, 93–105.

———. 1988. Late Chalcolithic and Early Bronze Age I Animal Remains from Arslantepe (Malatya): Preliminary Report. *Origini,* XII, 2a: 581–598.

———. 1988. *Analisi archeozoologica dello scheletro del cavallo nella necropoli di Vicenne.* Conoscenze. Campobasso, Riv. ann. d. Soprintend. Archeol. e per i Beni Ambient., Architett., Artist. e Stor. d. Molise, 4: 69–75.

———. 1988–1989. Takeover and Local Domestication: the Double-faced Nature of Early Animal Husbandry in South Italy. *Origini,* XIV: 371–386.

———. 1990. The Fauna (of Roccagloriosa). *In Roccagloriosa, I. L'abitato: scavo e ricognizione topografica (1976–1986),* eds. M. Gualtieri and H. Fracchia: . Naples, Bibliothéque de l'Institut Francais de Naples, Deuxieme Serie, V. VIII: 329–331.

———. 1990. Preliminary Report on the Animal Remains of Gabal Qutran (GQi) and Al-Masannah (MASi). In *The Bronze Age Culture of Hawlan at-Tiyal and Al-Hada,* ed. A. de Maigret. Rome, IsMEO, Centro Studi e Scavi Arch., Rep. and Mem. 24: 145–148.

———. 1991. The Mesolithic-Neolithic Transition in South Italy: The Switchover from Hunting to Animal Husbandry. In *Mésolithique et néolitisation en France et dans les régions limitrophes.* Actes du 113e Congrés National des Sociétés Savantes, Strasbourg, 5–9 avril 1988. Paris: 29–36.

———. 1991. L'allevamento. In *I Celti,* eds. S. Moscati et al. Milano 1991: 429–435.

———. 1992. Preliminary Information on the Faunal Remains from Excavations at Ras al-Junayz (Oman). In *South Asian Archaeology. Papers from the Tenth International Conference of South Asian Archaeologists in Western Europe,* Musée national des Arts asiatiques Guimet, Paris, France, 3–7 July 1989, ed. C. Jarrige. Madison, Wisc. Monogr. in World Archaeol., No. 14: 45–48.

———. 1992. The Possibilities of a Cooperation between Archaeology and Zoology. *Bulletino di Paletnologia Italiana* 83, NS I: 391–401.

———. 1993. Hunting in Arslantepe, Anatolia. In *Between the Rivers and Over the Mountains,* Archaeologia anatolica et mesopotamica Alba Palmieri dedicata, eds. Roma, M. Frangipane, H. Hauptmann, M. Liverani, P. Matthiae, M. Melling, 341–359.

———. 1993. I resti animali (di Castellar Rover). In *Castelar di Rover. Lo scavo di un castello medioevale,* Modena, Mat. d'Archeol., 2, ed. G. Rosada, 161–171.

———. 1993. The Sacrificial Remains from the Votive Deposit in F11. In *Fourth Century B.C. Magna Graecia: A Case Study.* Jonsered Studies in Mediterranean Archaeology and Literature 114, ed. M. Gualtieri, 125–127.

———. 1993. The Animal Remains (from Roccagloriosa). In *Fourth Century B.C. Magna Graecia: A Case Study.* Jonsered Studies in Mediterranean Archaeology and Literature 114, ed. M. Gualtieri, 281–290.

Bökönyi, S., and Siracusano, G. 1987. Reperti faunistici dell'Eta del Bronzo del sito di Coppa Nevigata: un commento preliminare. In *Coppa Nevigata e il suo territorio.* Testimonianze archeologiche dal VII al II millennio a C., eds. S. M. Cassano, A. Cazzella, A. Manfredini, M. Moscoloni. Roma: 205–210.

Bökönyi, S., and Bartosiewicz, L. 2000. A Review of Animal Remains from Shahr-i Sokhta (Eastern Iran). In *Archaeozoology of the Near East IVB,* Groningen, ARC Publication 32, eds. M. Mashkour, A. M. Choyke and H. Buitenhuis, 116–152.

Index

Numbers in italics refer to figures.